TRANSIT COOPERATIVE RESEARCH PROGRAM

TCRP REPORT 71

Track-Related Research

Volume 5:

Flange Climb Derailment Criteria
and Wheel/Rail Profile Management
and Maintenance Guidelines
for Transit Operations

HUIMIN WU
XINGGAO SHU
NICHOLAS WILSON
Transportation Technology Center, Inc. (TTCI)
Pueblo, CO

SUBJECT AREAS
Public Transit • Rail

Research Sponsored by the Federal Transit Administration in Cooperation with the Transit Development Corporation

TRANSPORTATION RESEARCH BOARD

WASHINGTON, D.C.
2005
www.TRB.org

TRANSIT COOPERATIVE RESEARCH PROGRAM

The nation's growth and the need to meet mobility, environmental, and energy objectives place demands on public transit systems. Current systems, some of which are old and in need of upgrading, must expand service area, increase service frequency, and improve efficiency to serve these demands. Research is necessary to solve operating problems, to adapt appropriate new technologies from other industries, and to introduce innovations into the transit industry. The Transit Cooperative Research Program (TCRP) serves as one of the principal means by which the transit industry can develop innovative near-term solutions to meet demands placed on it.

The need for TCRP was originally identified in *TRB Special Report 213—Research for Public Transit: New Directions*, published in 1987 and based on a study sponsored by the Urban Mass Transportation Administration—now the Federal Transit Administration (FTA). A report by the American Public Transportation Association (APTA), *Transportation 2000*, also recognized the need for local, problem-solving research. TCRP, modeled after the longstanding and successful National Cooperative Highway Research Program, undertakes research and other technical activities in response to the needs of transit service providers. The scope of TCRP includes a variety of transit research fields including planning, service configuration, equipment, facilities, operations, human resources, maintenance, policy, and administrative practices.

TCRP was established under FTA sponsorship in July 1992. Proposed by the U.S. Department of Transportation, TCRP was authorized as part of the Intermodal Surface Transportation Efficiency Act of 1991 (ISTEA). On May 13, 1992, a memorandum agreement outlining TCRP operating procedures was executed by the three cooperating organizations: FTA, The National Academies, acting through the Transportation Research Board (TRB); and the Transit Development Corporation, Inc. (TDC), a nonprofit educational and research organization established by APTA. TDC is responsible for forming the independent governing board, designated as the TCRP Oversight and Project Selection (TOPS) Committee.

Research problem statements for TCRP are solicited periodically but may be submitted to TRB by anyone at any time. It is the responsibility of the TOPS Committee to formulate the research program by identifying the highest priority projects. As part of the evaluation, the TOPS Committee defines funding levels and expected products.

Once selected, each project is assigned to an expert panel, appointed by the Transportation Research Board. The panels prepare project statements (requests for proposals), select contractors, and provide technical guidance and counsel throughout the life of the project. The process for developing research problem statements and selecting research agencies has been used by TRB in managing cooperative research programs since 1962. As in other TRB activities, TCRP project panels serve voluntarily without compensation.

Because research cannot have the desired impact if products fail to reach the intended audience, special emphasis is placed on disseminating TCRP results to the intended end users of the research: transit agencies, service providers, and suppliers. TRB provides a series of research reports, syntheses of transit practice, and other supporting material developed by TCRP research. APTA will arrange for workshops, training aids, field visits, and other activities to ensure that results are implemented by urban and rural transit industry practitioners.

The TCRP provides a forum where transit agencies can cooperatively address common operational problems. The TCRP results support and complement other ongoing transit research and training programs.

TCRP REPORT 71: Volume 5

Project D-7
ISSN 1073-4872
ISBN 0-309-08830-5
Library of Congress Control Number 2001135523

© 2005 Transportation Research Board

Price $25.00

NOTICE

The project that is the subject of this report was a part of the Transit Cooperative Research Program conducted by the Transportation Research Board with the approval of the Governing Board of the National Research Council. Such approval reflects the Governing Board's judgment that the project concerned is appropriate with respect to both the purposes and resources of the National Research Council.

The members of the technical advisory panel selected to monitor this project and to review this report were chosen for recognized scholarly competence and with due consideration for the balance of disciplines appropriate to the project. The opinions and conclusions expressed or implied are those of the research agency that performed the research, and while they have been accepted as appropriate by the technical panel, they are not necessarily those of the Transportation Research Board, the National Research Council, the Transit Development Corporation, or the Federal Transit Administration of the U.S. Department of Transportation.

Each report is reviewed and accepted for publication by the technical panel according to procedures established and monitored by the Transportation Research Board Executive Committee and the Governing Board of the National Research Council.

Special Notice

The Transportation Research Board of The National Academies, the National Research Council, the Transit Development Corporation, and the Federal Transit Administration (sponsor of the Transit Cooperative Research Program) do not endorse products or manufacturers. Trade or manufacturers' names appear herein solely because they are considered essential to the clarity and completeness of the project reporting.

Published reports of the

TRANSIT COOPERATIVE RESEARCH PROGRAM

are available from:

Transportation Research Board
Business Office
500 Fifth Street, NW
Washington, DC 20001

and can be ordered through the Internet at
http://www.national-academies.org/trb/bookstore

Printed in the United States of America

THE NATIONAL ACADEMIES
Advisers to the Nation on Science, Engineering, and Medicine

The **National Academy of Sciences** is a private, nonprofit, self-perpetuating society of distinguished scholars engaged in scientific and engineering research, dedicated to the furtherance of science and technology and to their use for the general welfare. On the authority of the charter granted to it by the Congress in 1863, the Academy has a mandate that requires it to advise the federal government on scientific and technical matters. Dr. Ralph J. Cicerone is president of the National Academy of Sciences.

The **National Academy of Engineering** was established in 1964, under the charter of the National Academy of Sciences, as a parallel organization of outstanding engineers. It is autonomous in its administration and in the selection of its members, sharing with the National Academy of Sciences the responsibility for advising the federal government. The National Academy of Engineering also sponsors engineering programs aimed at meeting national needs, encourages education and research, and recognizes the superior achievements of engineers. Dr. William A. Wulf is president of the National Academy of Engineering.

The **Institute of Medicine** was established in 1970 by the National Academy of Sciences to secure the services of eminent members of appropriate professions in the examination of policy matters pertaining to the health of the public. The Institute acts under the responsibility given to the National Academy of Sciences by its congressional charter to be an adviser to the federal government and, on its own initiative, to identify issues of medical care, research, and education. Dr. Harvey V. Fineberg is president of the Institute of Medicine.

The **National Research Council** was organized by the National Academy of Sciences in 1916 to associate the broad community of science and technology with the Academy's purposes of furthering knowledge and advising the federal government. Functioning in accordance with general policies determined by the Academy, the Council has become the principal operating agency of both the National Academy of Sciences and the National Academy of Engineering in providing services to the government, the public, and the scientific and engineering communities. The Council is administered jointly by both the Academies and the Institute of Medicine. Dr. Ralph J. Cicerone and Dr. William A. Wulf are chair and vice chair, respectively, of the National Research Council.

The **Transportation Research Board** is a division of the National Research Council, which serves the National Academy of Sciences and the National Academy of Engineering. The Board's mission is to promote innovation and progress in transportation through research. In an objective and interdisciplinary setting, the Board facilitates the sharing of information on transportation practice and policy by researchers and practitioners; stimulates research and offers research management services that promote technical excellence; provides expert advice on transportation policy and programs; and disseminates research results broadly and encourages their implementation. The Board's varied activities annually engage more than 5,000 engineers, scientists, and other transportation researchers and practitioners from the public and private sectors and academia, all of whom contribute their expertise in the public interest. The program is supported by state transportation departments, federal agencies including the component administrations of the U.S. Department of Transportation, and other organizations and individuals interested in the development of transportation. www.TRB.org

www.national-academies.org

COOPERATIVE RESEARCH PROGRAMS STAFF FOR TCRP REPORT 71, VOLUME 5

ROBERT J. REILLY, *Director, Cooperative Research Programs*
CHRISTOPHER W. JENKS, *TCRP Manager*
EILEEN P. DELANEY, *Director of Publications*
BETH HATCH, *Assistant Editor*

TCRP PROJECT D-7 PANEL
Field of Engineering of Fixed Facilities

ANTHONY BOHARA, *Southeastern Pennsylvania Transportation Authority, Philadelphia, PA* (Chair)
MICHAEL O. BROWN, *Bay Area Rapid Transit District, Oakland, CA*
STELIAN CANJEA, *New Jersey Transit Corporation, Bloomfield, NJ*
LANCE G. COOPER, *West Palm Beach, FL*
EARLE M. HUGHES, *Gannett Fleming Transit & Rail Systems, Audubon, PA*
JEFFREY G. MORA, *Washington, DC*
JAMES NELSON, *Wilson, Ihrig & Associates, Inc., Oakland, CA*
JOSEPH A. ORIOLO, *Massachusetts Bay Transportation Authority, Jamaica Plain, MA*
CHARLES L. STANFORD, *PB Transit & Rail Systems, San Francisco, CA*
TERRELL WILLIAMS, *FTA Liaison Representative*
LOUIS F. SANDERS, *APTA Liaison Representative*
GUNARS SPONS, *FRA Liaison Representative*
ELAINE KING, *TRB Liaison Representative*

FOREWORD

By Christopher W. Jenks
TCRP Manager
Transportation Research Board

This report includes the results of a research task carried out under TCRP Project D-7, "Joint Rail Transit-Related Research with the Association of American Railroads/Transportation Technology Center, Inc." The report includes flange climb derailment criteria for transit vehicles that include lateral-to-vertical (L/V) ratio limits and a corresponding flange-climb-distance limit, and it offers guidance that transit agencies can follow in their wheel and rail maintenance practices. This report should be of interest to engineers involved in the design, construction, maintenance, and operation of rail transit systems.

Over the years, a number of track-related research problem statements have been submitted for consideration in the TCRP project selection process. In many instances, the research requested has been similar to research currently being performed for the Federal Railroad Administration (FRA) and the freight railroads by the Transportation Technology Center, Inc. (TTCI), Pueblo, Colorado, a subsidiary of the Association of American Railroads (AAR). Transit track, signal, and rail vehicle experts reviewed the research being conducted by TTCI. Based on this effort, a number of research topics were identified where TCRP funding could be used to take advantage of research currently being performed at the TTCI for the benefit of the transit industry. A final report on one of these efforts—Flange Climb Derailment Criteria and Wheel/Rail Profile Management and Maintenance Guidelines for Transit Operations—is presented in this publication.

A railroad train running along a track is one of the most complex dynamic systems in engineering due to the presence of many nonlinear components. Wheel and rail geometries have a significant effect on vehicle dynamic performance and operating safety. The wheel/rail interaction in transit operations has its own special characteristics. Transit systems have adopted different wheel and rail profile standards for different reasons. Older systems with long histories have wheel and rail profile standards that were established many years ago. Newer systems have generally selected wheel and rail profiles based on an increased understanding of wheel/rail interaction in recent years.

Transit systems are typically operated in dense urban areas, which frequently results in systems that contain a large number of curves with small radii that can increase wheel and rail wear and increase the potential for flange-climb derailments. Transit systems also operate a wide range of vehicle types, such as those used in commuter rail, light rail, and rapid transit services, with a wide range of suspension designs and performance characteristics. Increasing operating speed and the introduction of new vehicle designs have posed an even greater challenge for transit systems to maintain and improve wheel/rail interaction.

Under TCRP Project D-7 Task 8, TTCI was asked to develop flange climb derailment criteria derived from wheel profiles found in various types of transit vehicles. In ddition, TTCI was asked to develop guidelines for the maintenance and management of wheel/rail profiles for transit vehicles. In meeting these objectives, TTCI first identified common problems and concerns related to wheel/rail profiles through a survey of representative transit systems. Based on this information, flange climb derailment criteria were developed using wheel profiles identified during the survey. TTCI then validated the flange climb derailment criteria using test track data and computer simulation. Finally, TTCI developed guidelines for the management and maintenance of wheel/rail profiles for transit operations based on problems and concerns identified during the transit agency survey and current transit practice.

CONTENTS

1 SUMMARY

3 CHAPTER 1 Introduction
  1.1 Background, 3
  1.2 Structure of This Report, 3
  1.3 Summary of Phase I Work, 4

6 CHAPTER 2 Flange Climb Derailment Criteria
  2.1 Wheel L/V Ratio Criteria, 6
  2.2 Flange-Climb-Distance Criteria, 6
  2.3 Determination of Effective AOA, 7
  2.4 Definition of Flange Climb Distance, 8
  2.5 A Biparameter Technique to Derive Flange Climb Distance, 9
  2.6 Effect of Speed on Distance to Climb, 10
  2.7 Application of Flange Climb Criteria, 10
  2.8 Examples of Application of Flange Climb Criteria, 10

12 CHAPTER 3 Recommended Management and Maintenance Guidelines of Wheel/Rail Profiles for Transit Operations
  3.1 Requirement for New Wheel Profile Drawings, 12
  3.2 Wheel/Rail Profile Measurement and Documentation, 14
  3.3 Wheel/Rail Profile Assessment, 16
  3.4 Understanding Important Stages of Wheel/Rail Contact in a System, 23
  3.5 Wheel Re-Profiling, 24
  3.6 Wheel Profile Design, 25
  3.7 Ground Rail Profile, 26
  3.8 Effect of Gage and Flange Clearance on Wheel/Rail Contact, 28
  3.9 Wheel/Rail Profile Monitoring Program, 30

31 CHAPTER 4 Glossary of Technical Terms

33 REFERENCES

A-1 APPENDIX A Effect of Wheel/Rail Profiles and Wheel/Rail Interaction on System Performance and Maintenance in Transit Operations

B-1 APPENDIX B Investigation of Wheel Flange Climb Derailment Criteria for Transit Vehicles (Phase I Report)

C-1 APPENDIX C Investigation of Wheel Flange Climb Derailment Criteria for Transit Vehicles (Phase II Report)

FLANGE CLIMB DERAILMENT CRITERIA AND WHEEL/RAIL PROFILE MANAGEMENT AND MAINTENANCE GUIDELINES FOR TRANSIT OPERATIONS

SUMMARY   The objective of this research was to improve wheel/rail interaction in transit systems by introducing flange climb derailment criteria and wheel/rail profile management and maintenance guidelines that can be applied to transit operations.

This work was started with a survey conducted on six representative transit systems to define the common problems and concerns related to wheel/rail profiles in transit operation. As an integral part of the survey, the research team provided wheel/rail interaction training seminars to maintenance and engineering staff at each of the systems. The survey results are compiled in Appendix A of this report, and Appendices B and C discuss the development of flange climb derailment criteria for transit vehicles.

The flange climb derailment criteria developed include a wheel lateral-to-vertical (L/V) ratio limit and a corresponding flange-climb-distance limit. These criteria were developed based on computer simulations of single wheelsets, and representative transit vehicles. The resulting criteria are shown to be dependent on wheel/rail contact angle, wheel/rail friction coefficient, flange length, and wheelset angle of attack (AOA).

The wheel profiles used in the simulations were obtained from the transit system survey. These profiles were applied to simulations of both light rail and rapid transit vehicles with flange angles ranging from 60 degrees to 75 degrees and flange length ranging from 0.395 to 0.754 in.

The proposed criteria were validated using flange-climb test data collected with the research team Track Loading Vehicle (TLV). An example of applying the criteria to a passenger car test is given in the report. The limitations of the proposed criteria are also discussed.

A general form of flange-climb-distance criterion is proposed in this report. It applies to an L/V ratio equal to or less than 1.99. A biparameter regression technique was developed to derive the distance criterion, which is more accurate and less conservative, but only for the specific wheel and rail simulated.

Because of the wide diversity of practices currently applied, it is not possible to set universal rules that can be applied to all transit systems. However, it is beneficial to recommend general guidelines that transit operations can follow in their wheel and rail maintenance practice. Therefore, some guidelines in this report are rather more conceptive than quantitative. Guidelines and recommendations applying to the management and maintenance of wheel/rail profiles for transit operations involving the following areas have been provided:

- New wheel profile drawings
- Wheel/rail profile measurement and documentation
- Wheel/rail profile assessment
- Wheel re-profiling
- Wheel profile design
- Ground rail profile design
- Effect of gage and flange clearance on wheel/rail interaction
- Wheel/rail profile monitoring program

CHAPTER 1

INTRODUCTION

The objective of this project was to improve wheel/rail interaction in transit systems by introducing flange climb derailment criteria and wheel/rail profile management and maintenance guidelines that can be applied to transit operations.

1.1 BACKGROUND

A railroad train running along a track is one of the most complex dynamic systems in engineering due to the many nonlinear components in the system. In particular, the interaction between wheel and rail is a very complicated nonlinear element in the railway system. Wheel and rail geometries, involving both cross sectional profiles and geometry along the moving direction with varying shapes due to wear, have a significant effect on vehicle dynamic performance and operating safety.

The wheel/rail interaction in transit operations has its own special characteristics. Without the requirement for interoperability, transit systems have adopted different wheel and rail profile standards for different reasons. Some of these standards are unique to a particular system. Older systems with long histories frequently have wheel and rail profile standards that were established many years ago. For some older systems, the reasons that specific profiles were adopted have been lost in time. Newer systems have generally selected the wheel/rail profiles based on the increased understanding of wheel/rail interaction in recent years.

Transit systems are usually operated in dense urban areas, which frequently results in lines that contain a large percentage of curves or curves with small radii, which can increase wheel and rail wear and increase the potential for flange climb derailments. Transit systems also operate a wide range of vehicle types—such as those used in commuter rail service, heavy or rapid transit, and light rail vehicles—with a wide range of suspension designs and performance characteristics. Increasing operating speed and introducing new designs of vehicles have posed an even greater challenge for transit systems to maintain and improve wheel/rail interaction.

Considering the special features of transit operations, the purpose of this report is first to propose a general form of flange climb derailment criterion derived from wheel profiles applied in transit vehicles and, second, to provide the guidelines for applying the management and maintenance of wheel/rail profiles for transit operators.

Due to the diversity among practices currently applied by different rail transit systems, it is not possible to set universal rules that can be applied to all transit systems. However, it will be beneficial to recommend certain guidelines that transit operations can follow in their wheel and rail maintenance practice. Thus, some guidelines in this report are rather more conceptive than quantitative.

The contents of this report were compiled as the result of a review of the literature pertaining to flange climb derailment, wheel/rail interaction and wheel/rail profiles; extensive research on the development of flange climb derailment criteria; the investigation on several representative transit systems; and the authors' experiences on a number of projects previously conducted in related fields.

1.2 STRUCTURE OF THIS REPORT

This work was performed in two phases. In Phase I, the common problems and concerns related to wheel/rail profiles were defined through a survey conducted on representative transit systems. The preliminary flange climb derailment criteria, derived using the wheel profiles collected in the transit system survey, were proposed. Two reports were produced after the Phase I work and are presented as appendices:

- Appendix A: Effect of Wheel/Rail Profiles and Wheel/Rail Interaction on System Performance and Maintenance in Transit Operations, and
- Appendix B: Investigation of Wheel Flange Climb Derailment Criteria for Transit Vehicles (Phase I Report).

Section 1.3 of this report briefly summarizes the results of Phase I.

In Phase II, the flange climb derailment criteria developed in the Phase I work were further validated by track test data. A general form of the criterion is proposed in this report. A report detailing the validation and development of the criteria, "Investigation of Wheel Flange Climb Derailment Criteria for Transit Vehicles (Phase II Report)," is attached as Appendix C.

The validated flange climb criteria are stated in Chapter 2 of this report with examples of applications in simulation and track test for evaluating flange climb derailment.

In Phase II, the guidelines for applying to management and maintenance of wheel/rail profiles for transit operations were recommended based on the problems and concerns uncovered in the survey and current transit operations practices. Chapter 3 presents these guidelines.

Chapter 4 is a glossary provided to help the reader better understand the technical terms used in this report relating to the flange climb criteria and the wheel/rail profile.

1.3 SUMMARY OF PHASE I WORK

Onsite surveys were conducted at six representative transit systems to investigate current practices and concerns related to wheel/rail profiles in transit operations. After compiling the information received and analyzing wheel and rail profiles that were collected onsite, common problems and concerns related to wheel/rail profiles and wheel/rail interaction were identified. These are summarized below:

- Adoption of low wheel flange angles can increase the risk of flange climb derailment. High flange angles above 72 degrees are strongly recommended to improve operational safety.
- Rough surfaces from wheel truing can increase the risk of flange climb derailment. Smoothing the surface after wheel truing and lubrication could mitigate the problem considerably.
- Introduction of new wheel and rail profiles needs to be carefully programmed for both wheel truing and rail grinding to achieve a smooth transition from old wheel/rail profiles to new profiles.
- Independently rotating wheels can produce higher lateral forces and higher wheel/rail wear on curves without adequate control mechanisms.
- Cylindrical wheels may reduce the risk of vehicle hunting (lateral instability), but can have poor steering performance on curves.
- Some wheel and rail profile combinations used in transit operations were not systematically evaluated to ensure that they have good performance on both tangent track and curves under given vehicle and track conditions.
- Severe two-point contact has been observed on the designed wheel/rail profile combinations at several transit operations. This type of contact tends to produce poor steering on curves resulting in higher lateral force and higher rate of wheel/rail wear.
- Track gage and restraining rails need to be carefully set on curves to allow sufficient usage of rolling radius difference (RRD) generated and to mitigate high rail wear and lateral force.
- Wheel slide and wheel flats are an issue for almost every system, especially during the fall season. Although several technologies have been applied to lessen the problem, more effective ones are needed.
- Noise related to wheels and rails generally are caused by wheel screech/squeal, wheel impact, and rail corrugations. Lubrication and optimizing wheel/rail contact would help to mitigate these problems.
- Friction management is a field that needs to be further explored. Application of lubrication is very limited in transit due to the complications related to wheel slide and wheel flats.
- Reduction of wheel/rail wear can be achieved by optimization of wheel/rail profiles, properly designed truck primary suspension, improvement of track maintenance, and application of lubrication.
- Without a wheel/rail profile measurement and documentation program, transit operators will have difficulty reaching a high level of effectiveness and efficiency in wheel/rail operation and maintenance.
- Further improvement of transit system personnel understanding of wheel/rail profiles and interaction should be one of the strategic steps in system improvement. With better understanding of the basic concepts, vehicle/track operation and maintenance would be performed more effectively.

Appendix A provides further information about the above issues.

An investigation of wheel flange climb derailment criteria as applied to transit operation was conducted by extensive computer simulations using the wheel/rail profiles collected from several transit systems. Based on simulations of single wheelsets, preliminary lateral-to-vertical (L/V) ratio and climb-distance criteria for transit vehicle wheelsets were proposed. The proposed criteria were further validated through simulation of three types of transit vehicles. This research has been based on the methods previously used by the research team to develop flange climb derailment criteria for the North American freight railroads. The main conclusions drawn from this study are summarized below:

- The single wheel L/V ratio required for flange climb derailment is determined by the wheel maximum flange angle, friction coefficient, and wheelset AOA.
- The L/V ratio required for flange climb converges to Nadal's value for AOA greater than 10 milliradians (mrad). For lower wheelset AOA, the wheel L/V ratio necessary for flange climb becomes progressively higher than Nadal's value.
- The distance required for flange climb derailment is determined by the L/V ratio, wheel maximum flange angle, wheel flange length, and wheelset AOA.
- The flange-climb distance converges to a limiting value at higher AOA and higher L/V ratios. This limiting

value strongly correlates with wheel flange length. The longer the flange length, the longer the climb distance. For lower wheelset AOA, when the L/V ratio is high enough for the wheel to climb, the wheel-climb distance for derailment becomes progressively longer than the proposed flange-climb-distance limit. The wheel-climb distance at lower wheelset AOA is mainly determined by the maximum flange angle and L/V ratio.
- Besides the flange contact angle, flange length also plays an important role in preventing derailment. The climb distance can be increased through use of higher wheel maximum flange angles and longer flange length.
- The flanging wheel friction coefficient significantly affects the wheel L/V ratio required for flange climb. The lower the friction coefficient, the higher the single wheel L/V ratio required.
- For conventional solid wheelsets, a low, nonflanging wheel friction coefficient has a tendency to cause flange climb at a lower flanging wheel L/V ratio. Flange climb occurs over a shorter distance for the same flanging wheel L/V ratio.
- For independently rotating wheelsets, the effect of non-flanging wheel friction coefficient is negligible because the longitudinal creep force vanishes.
- Increasing vehicle speed increases the distance to climb.

The Phase I report detailing this study of flange climb criteria is given in Appendix B.

CHAPTER 2

FLANGE CLIMB DERAILMENT CRITERIA

The flange climb derailment criteria proposed in this section include the wheel L/V ratio limit and the flange-climb-distance limit. Details of the research to develop these criteria are reported in Appendices B and C. These criteria were developed based on computer simulations of single wheelsets. The wheel profiles used in the simulations were obtained from the transit system survey. These profiles were applied on both light rail and rapid transit vehicles with flange angles ranging from 60 degrees to 75 degrees and flange lengths ranging from 0.395 to 0.754 in.

The proposed criteria have been validated by flange-climb test data using the TLV. This section provides an example of applying the criteria to a passenger car test. The limitations of the proposed criteria are also discussed.

2.1 WHEEL L/V RATIO CRITERIA

The L/V ratio criteria proposed for transit vehicles are stated as follows:

(1) if AOA ≥ 5 mrad or if AOA is unknown,

$$\frac{L}{V} < q_0, \quad (2.1)$$

(2) If AOA < 5 mrad

$$\frac{L}{V} < q_0 + \frac{0.43}{AOA + 1.2}, \quad (2.2)$$

where q_0 is the Nadal value that is defined by Equation 2.3 and Figure 2.1 and AOA is wheelset AOA in mrad.

The Nadal single-wheel L/V limit criterion (1), proposed by M. J. Nadal in 1908 for the French Railways, has been used throughout the railroad community. Nadal proposed a limiting criterion as a ratio of L/V forces:

$$\frac{L}{V} = \frac{\tan(\delta) - \mu}{1 + \mu \tan(\delta)} \quad (2.3)$$

where μ is the friction coefficient at the wheel/rail contact surface and δ is the wheel/rail contact angle.

Figure 2.1 shows the Nadal values for different wheel/rail maximum contact angles and friction coefficient combinations. The Nadal values for contact angles of 63 and 75 degrees are specified in Figure 2.1 with values of 0.73 and 1.13, respectively. If the maximum contact angle is used, Equation 2.3 gives the minimum wheel L/V ratio at which flange climb derailment may occur for the given contact angle and friction coefficient μ. Clearly, wheels with low flange angles and high friction coefficient have a low L/V ratio limit and a higher risk of flange climb derailment.

Equation 2.1 states that if the AOA is larger than 5 mrad or the AOA cannot be determined (which is usually the case during on-track tests), the limiting L/V value is the Nadal value determined by Equation 2.3. If the AOA can be determined with an AOA measurement device or from simulation results and its value is less than 5 mrad, the limiting L/V ratio can be less conservative than the Nadal value (Equation 2.2). Equation 2.2 was developed to account for the effects of increased flange climb L/V with small AOAs. Figure 2.2 shows a comparison of Equation 2.2 and the Nadal value for a wheel with a 63 degree flange angle.

The study included in Appendix B indicates that one reason independently rotating wheelsets (IRW) tend to climb the rail more easily than conventional solid wheelsets is that the coefficient of friction on nonflanging wheel has no effect on the flanging wheel. Therefore, the Nadal L/V limit is accurate for IRW but can be conservative for the wheelsets of solid axles. The wheel L/V ratio required for flange climb for solid axles increases as the increased friction coefficient on nonflanging wheel. If the friction coefficient on the nonflanging wheel approaches to zero, the L/V ratio limit for the solid axle wheel would be the same as that for IRW.

2.2 FLANGE-CLIMB-DISTANCE CRITERIA

In practice, a flange climbing derailment is not instantaneous. The L/V ratio has to be maintained while the climbing takes place. If, for example, the lateral force returns to zero before the flange has reached the top of the rail, the wheel might be expected to drop down again. When the flange contacts the rail for a short duration, as may be the case during hunting (kinematic oscillations) of the wheelset, the L/V ratio might exceed Nadal's limit without flange climbing. For that reason, the flange-climb-distance criteria were developed to evaluate the risk of derailment associated

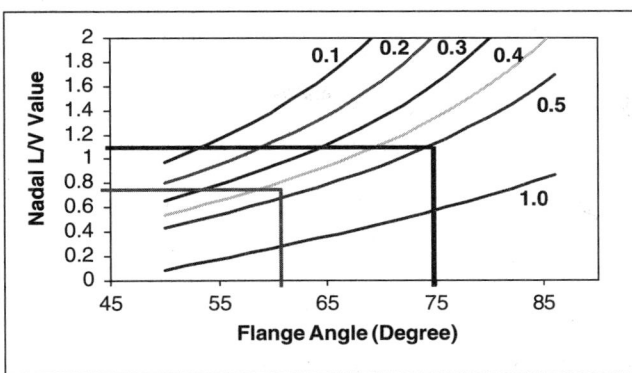

Figure 2.1. Nadal criterion as a function of flange angle and friction coefficients in a range of 0.1 to 1.0.

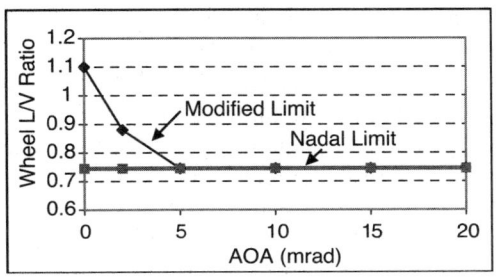

Figure 2.2. Comparison of Nadal L/V ratio limit and modified L/V ratio limit.

with the wheel L/V ratio limit. Flange climb derailment would occur only if both wheel L/V ratio limit and distance limit are exceeded.

A general form of flange-climb-distance criterion is proposed in this section that applies to an L/V ratio equal to or less than 1.99.

2.2.1 A General Flange-Climb-Distance Criterion

A general flange-climb-distance criterion was developed by using the technique described in Appendix C of this report. Sixteen combinations of wheel flange angle and flange length, covering a wide range of these values on actual wheels, were used for the derivation. This general criterion, proposed in Equation 2.4, takes the AOA, the maximum flange angle, and flange length as parameters.

$$D < \frac{A * B * Len}{AOA + B * Len} \quad (2.4)$$

where D is limiting climb distance in feet and AOA is in mrad. Coefficients A, B are functions of the maximum flange angle Ang (degrees) and flange length Len (in.) as defined in Section 2.4:

$$A = \left(\frac{100}{-1.9128\,Ang + 146.56} + 3.1\right) *$$

$$Len - \frac{1}{-0.0092(Ang)^2 + 1.2152\,Ang - 39.031} + 1.23$$

$$B = \left(\frac{10}{-21.157\,Len + 2.1052} + 0.05\right) *$$

$$Ang + \frac{10}{0.2688\,Len - 0.0266} - 5$$

The limiting climb distance for a specific transit wheel profile can be derived from the above general criterion by substituting the maximum flange angle and flange length into Equation 2.4. It is especially useful for the transit wheel profiles that were not simulated in this report.

Table 2.1 lists a range of limiting flange-climb-distance values computed using Equation 2.4 for a specified range of flange angles, flange length, and AOA. Table 2.1 indicates that at an AOA of 5 mrad, the limiting flange-climb distance increases as increased wheel flange angle and flange length. At an AOA of 10 mrad, flange length has more effect on the distance limit than flange angle.

In summary, considering that flange climb generally occurs at a higher AOA, increasing wheel flange angle can increase the wheel L/V ratio limit required for flange climb and increasing flange length can increase the limiting flange climb distance.

2.3 DETERMINATION OF EFFECTIVE AOA

The flange climb criteria, including both wheel L/V ratio limit and climbing distance limit, are closely related to the wheelset AOA. Because the wheelset AOA may not be

TABLE 2.1 Limiting flange-climb-distance computed using Equation 2.4

	AOA = 5 mrad				AOA = 10 mrad			
Flange Angle (deg)	63 deg	68 deg	72 deg	75 deg	63 deg	68 deg	72 deg	75 deg
Flange Length (inch)								
0.4 inch	2.0	2.2	2.4	3.3	1.5	1.5	1.5	1.9
0.52 inch	2.4	2.6	2.9	3.8	1.8	1.8	1.8	2.1
0.75 inch	3.2	3.5	3.7	4.3	2.3	2.3	2.2	2.4

TABLE 2.2 Estimation of constant c

Vehicle Type	Axle Spacing Distance (in.)	Constant c
LRV1 (with IRW)	74.8	3.08
LRV2 (Solid axles)	75	2.86
HRV (Solid axles)	82	2.04

TABLE 2.3 Conservative AOAe (mrad) for practical use

Vehicle and Truck Type	Straight Lines	5-Degree Curves	10-Degree Curves	>10-Degree Curves
Vehicle with IRW	10	15	20	Equation 2.5 + 10
Others	5	10	15	Equation 2.5 + 5

available or cannot be measured under certain circumstances, an equivalent index, or effective AOA (AOAe), is proposed here in order to use the flange climb criteria. The AOAe is a function of axle spacing in the truck, track curvature, and truck type.

The equivalent index AOAe (in milliradians) of the leading axle of a two-axle truck can be obtained through a geometric analysis of truck geometry on a curve (Equation 2.5):

$$AOAe = 0.007272clC = \frac{41.67cl}{R} \qquad (2.5)$$

where

c = a constant for different truck types,
l = axle spacing distance (in.),
C = curve curvature (degrees), and
R = curve radius (ft).

Table 2.2 lists the constant c obtained from simulations for three types of representative transit vehicles: a Light Rail Vehicle Model 1 (LRV1) with independent rolling wheels in the center truck, a Light Rail Vehicle Model 2 (LRV2), and a Rapid Transit Vehicle (HRV). Therefore, the AOAe can be estimated according to the track curvature (C) and known constant c (Equation 2.5).

Due to the track perturbations and the degrading of wheelset steering capability, the practical wheelset AOA could be higher than the value calculated by Equation 2.5. Table 2.3 shows the AOAe values recommended for use in the distance criterion of Equation 2.4. These values for AOAe were considered conservative enough according to the simulation results and test data.

When the vehicle runs on a curve with the curvature lower than 10 degrees and not listed in Table 2.3, it is recommended that a linear interpolation between the segment points in Table 2.3 should be used in the criterion, as shown in Figure 2.3. The statistical data from an AOA wayside monitoring system should be used in the criterion to take into account the many factors affecting AOAe if such a system is available.

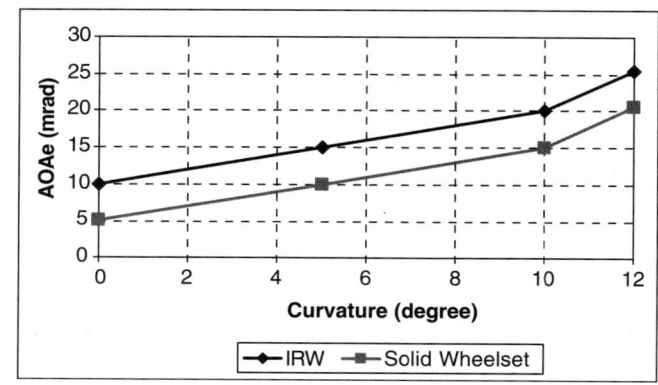

Figure 2.3. Recommended conservative AOAe for practical use.

2.4 DEFINITION OF FLANGE CLIMB DISTANCE

The climb distance used here is defined as the distance traveled starting from the point at which the limiting L/V ratio (Equation 2.1 and 2.2) is exceeded (equivalent to the point "A" in Figure B-4 of Appendix B) to the point of derailment. For the purposes of these studies, the point of derailment was determined by the contact angle on the flange tip decreasing to 26.6 degrees after passing the maximum contact angle.

The 26.6-degree contact angle corresponds to the minimum contact angle for a friction coefficient of 0.5. Figure 2.4 shows the wheel flange tip in contact with the rail at a 26.6-degree angle. Between the maximum contact angle (point Q) and the 26.6-degree flange tip angle (point O), the wheel can slip back down the gage face of the rail due to its own vertical axle load if the external lateral force is suddenly reduced to zero. In this condition, the lateral creep force F (due to AOA) by itself is not large enough to cause the wheel to derail.

When the wheel climbs past the 26.6-degree contact angle (point O) on the flange tip, the wheel cannot slip back down the gage face of the rail due to its own vertical axle load: the lateral creep force F generated by the wheelset AOA is large

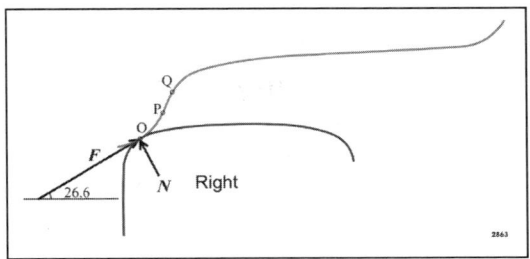

Figure 2.4. Wheel/rail interaction and contact forces on flange tip.

enough to resist the fall of the wheel and force the flange tip to climb on top of the rail.

The flange length Len is defined as the sum of the maximum flange angle arc length QP and flange tip arc length PO, as shown in Figure 2.4.

2.5 A BIPARAMETER TECHNIQUE TO DERIVE FLANGE CLIMB DISTANCE

A biparameter regression technique was also developed to derive the distance criterion. The limiting distances derived from the biparameter regression technique are more accurate and less conservative than that defined by the general form of distance criterion presented in Section 2.2.1. However, the derivation must be performed for each specific wheel and rail profile combination. An example of derivation of the distance criterion, using the biparameter method for the AAR-1B wheel contacting AREMA 136 RE rail, is demonstrated in this section.

In Appendix C of this report, the bilinear characteristic between the transformed climb distance and the two parameters, AOA and L/V ratio, was obtained through a nonlinear transformation. The accuracy of the fitting formula is further improved by using gradual linearization methodology.

As an example of this technique, a biparameter flange-climb-distance criterion, which takes the AOA, the L/V ratio as parameters, was proposed for vehicles with AAR-1B wheel and AREMA 136-pound rail profile:

$$\text{L/V Distance (feet)} < \frac{1}{0.001411 * AOA + (0.0118 * AOA + 0.1155) * L/V - 0.0671} \quad (2.6)$$

The biparameter criterion has been validated by the TLV test data. Some application limitations of the biparameter criterion (Equation 2.6) include the following:

- The L/V ratio in the biparameter criterion must be higher than the L/V limit ratio corresponding to the AOA, because no flange climb can occur if the L/V ratio is lower than the limit ratio.
- The biparameters criterion is obtained by fitting in the bilinear data range where AOA is larger than 5 mrad. It is conservative at AOA less than 5 mrad due to the non-linear characteristic.
- The biparameter criterion was derived based on simulation results for the new AAR-1B wheel on new AREMA 136-pound rail. It is only valid for vehicles with this combination of wheel and rail profiles.

Figure 2.5 shows the limiting flange-climb distance defined by the general form of flange-climb-distance criterion (Equation 2.4) compared to the biparameter criterion (Equation 2.6) for the combination of AAR-1B wheel and AREMA 136 RE rail at 10 mrad AOA. The AAR-1B wheel has a flange angle of 75 degrees and a flange length of 0.618 in.

Under this condition, the limiting flange-climb-distance given from the general form of climb-distance criterion is a constant of 2.3 ft once the wheel L/V ratio exceeds the Nadal limit of 1.13 for a friction coefficient of 0.5, as shown by the straight line in Figure 2.5.

The curve in Figure 2.5 gives the limiting distance criterion from the biparameter criterion under the same condition. Once the Nadal L/V ratio is exceeded, the distance limit is the function of the average of actual L/V ratios over the distance that Nadal L/V ratio limit has been exceeded. It can be seen from Figure 2.5 that the biparameter criterion is less conservative than the general form of distance criterion for wheel L/V ratio less than 2.0, especially when the wheel L/V ratio is just above the Nadal limit (1.13 in this case). In actual tests, sustained wheel L/V ratios greater than 2.0 are uncommon.

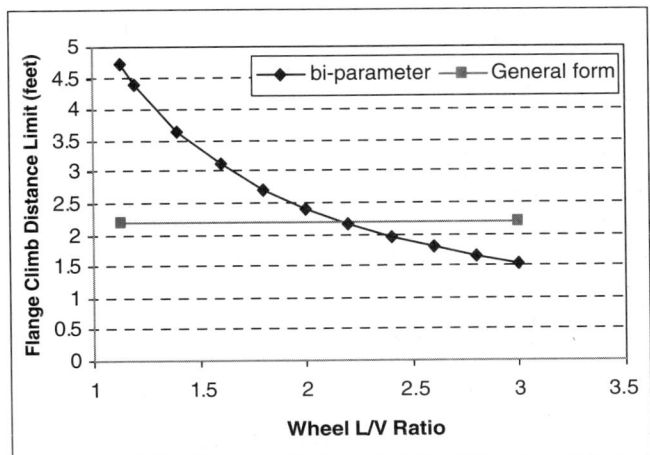

Figure 2.5. Comparison of flange-climb-distance limit from the general form of distance criterion and the biparameter criterion.

2.6 EFFECT OF SPEED ON DISTANCE TO CLIMB

The above criteria, both the general formula and biparameter method, were derived based on the flange climb simulation results of a single wheelset running at a speed of 5 mph. Simulation results show the climb distance slightly increases with increasing running speed due to increased longitudinal creep force and reduced lateral creep force (2), as shown in Figure 2.6.

The dynamic behavior of wheelset becomes very complicated at higher running speed (above 80 mph for 5 mrad AOA and above 50 mph for 10 mrad AOA). However, the distance limit derived from the speed of 5 mph should be conservative for higher operating speeds.

2.7 APPLICATION OF FLANGE CLIMB CRITERIA

2.7.1 In Simulations

The application of flange climb criteria in simulations can be found in Chapter 3 of Appendix B.

2.7.2 In Track Tests

In tests, when AOA is unknown or can't be measured, the AOAe described in Section 2.3 has to be estimated using Equation 2.8. The examples in the following section demonstrate the application of flange climb criteria in track tests.

2.8 EXAMPLES OF APPLICATION OF FLANGE CLIMB CRITERIA

As an example of their application, the flange climb criteria were applied to a passenger car with an H-frame truck undergoing dynamic performance tests at the FRA's Transportation Technology Center, Pueblo, Colorado, on July 28, 1997. The car was running at 20 mph through a 5 degree curve with 2 in. vertical dips on the outside rail of the curve. The L/V ratios were calculated from vertical and lateral forces measured from the instrumented wheelsets on the car.

Table 2.4 lists the 4 runs with L/V ratios higher than 1.13, exceeding the AAR Chapter XI flange climb safety criterion. The rails during the tests were dry, with an estimated friction coefficient of 0.5. The wheel flange angle was 75 degrees, resulting in a corresponding Nadal value of 1.13.

The climb distance and average L/V in Table 2.4 were calculated for each run from the point where the L/V ratio exceeded 1.13.

2.8.1 Application of General Flange Climb Criterion

The instrumented wheelset has the AAR-1B wheel profile with 75.13 degree maximum flange angle and 0.62 in. flange length; by substituting these two parameters into the general flange climb criterion, the flange criterion for the AAR-1B wheel profile is as follows:

$$D < \frac{26.33}{AOAe + 1.2}$$

The axle spacing distance for this rail car is 102 in. The constant c was adopted as 2.04 since the vehicle and truck design is similar to the heavy rail vehicle in Table 2.2. According to Equation 2.5, the AOAe is about 7.6 mrad for this passenger H-frame truck on a 5-degree curve. By substituting the AOAe into the above criteria, the safe climb distance without derailment is 3 ft. According to Table 2.3, the conservative AOAe for a 5-degree curve should be 10 mrad;

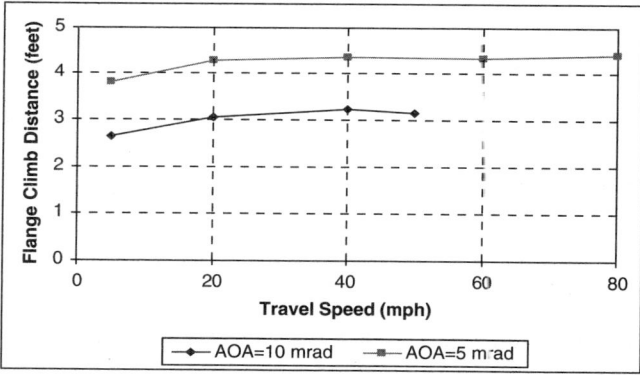

Figure 2.6. Effect of travel speed on distance to wheel climb. (L/V ratio = 1.99, AAR-1B wheel (75-degree flange angle) and AREMA 136 RE rail.)

TABLE 2.4 Passenger car test results: Climb distance and average L/V measured from the point where the L/V ratio exceeded 1.13, for friction coefficient of 0.5

Runs	Speed	Maximum L/V Ratio	Average L/V Ratio	Climb Distance
rn023	20.39 mph	1.79	1.39	5.8 ft
rn025	19.83 mph	2.00	1.45	6.3 ft
rn045	19.27 mph	1.32	1.23	0.7 ft
rn047	21.45 mph	1.85	1.52	5 ft

the conservative safe climb distance without derailment is 2.4 ft; however, the climb distance according to the 50 ms criterion is 1.4 ft. (The 50-ms criterion is discussed in Appendix B, Section B1.3.)

The wheel, which climbed 0.7 ft distance in run rn045 with a 1.23 average L/V ratio (maximum L/V ratio 1.32), was running safely without threat of derailment according to the criterion. The other three runs were unsafe because their climb distances exceeded the criterion.

2.8.2 Application of Biparameters Criterion

Figure 2.7 shows the application of the biparameters criterion on the same passenger car test. The run (rn045) with the maximum 1.32 L/V ratio is safe, since its climb distance of 0.7 ft is shorter than the 4.3-ft criterion value calculated by the biparameter formula (Equation 2.7). The climb distance is even below the 20 mrad AOAe criterion line, which seldom happened for an H-frame truck running on the 5-degree curve. The other three runs were running unsafely because their climb distances exceeded the 10 mrad conservative AOAe criterion line.

The same conclusion is drawn by applying the general flange climb criterion and the biparameter flange climb criterion to the passenger car test. The climb distances of these two criteria also show that the general flange climb criterion is more conservative than the biparameter criterion. The reason for this is that the average L/V ratio in the test, which is 1.23, is lower than the 1.99 ratio used in the simulation to derive the general flange climb criterion. The difference between these two criteria shows the biparameter flange climb criterion is able to reflect the variation of the L/V ratio. However, the general flange climb criterion is conservative for most cases since the sustained average 1.99 L/V ratio during flange climb is rare in practice.

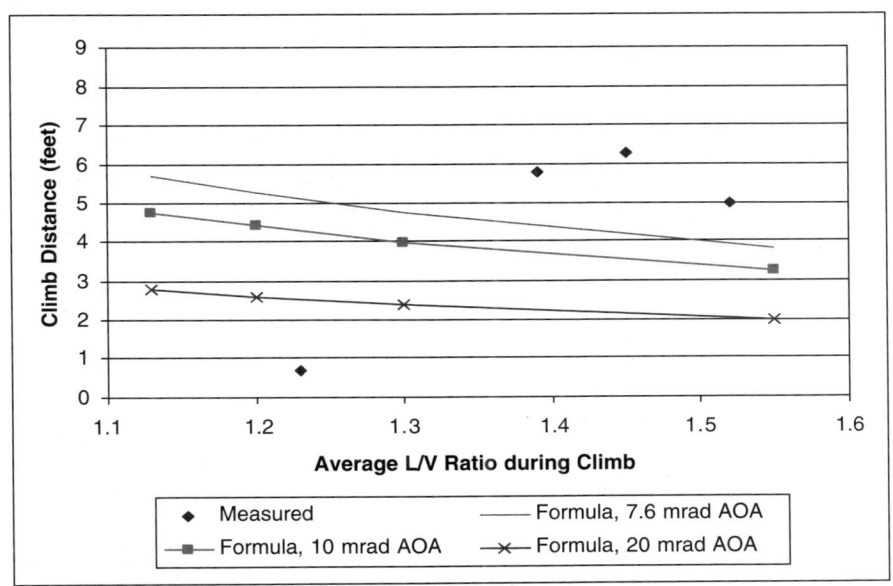

Figure 2.7. Application of the biparameter criterion for friction coefficient of 0.5.

CHAPTER 3

RECOMMENDED MANAGEMENT AND MAINTENANCE GUIDELINES OF WHEEL/RAIL PROFILES FOR TRANSIT OPERATIONS

In this section, the guidelines for management and maintenance of wheel/rail profiles for transit operations are recommended. The guidelines cover the following issues:

- New wheel profile drawings
- Wheel/rail profile measurement and documentation
- Wheel/rail profile assessment
- Wheel re-profiling
- Wheel profile design
- Ground rail profile design
- Effect of gage and flange clearance on wheel/rail interaction
- Wheel/rail profile monitoring program

In order to better understand and apply the guidelines, related technical definitions are briefly introduced. Specific techniques (or software) mentioned in this report are only used as examples and do not indicate endorsement of specific products.

3.1 REQUIREMENT FOR NEW WHEEL PROFILE DRAWINGS

Wheel profile drawings represent the designed shapes for new wheels. Adoption of a wheel profile design for a specific transit operation requires careful consideration of the vehicle and track conditions that the new wheel profile will experience. When building a new system or a new line in particular, selecting a proper wheel profile at the start is very important for the long-term stability of the system, which is indicated by good vehicle performance and low-wheel/rail-wear rates.

Therefore, the requirements for the wheel profile drawings, described below, are not only for manufacturer use but also for wheel profile designers to recognize the important parameters, and for staff in transit operation and maintenance to understand the features of the wheel profile(s) used in their system.

For the purpose of wheel manufacturing, wheel profile drawings generally have all dimension descriptions required for the machine production of such a profile. In this section, only those parameters that currently are not shown or not required on the drawing are emphasized. They are the following:

- Wheel flange angle
- Wheel flange length
- Wheel tread taper
- Coordinates of wheel profile

3.1.1 Wheel Flange Angle

Wheel flange angle is defined as the maximum angle of the wheel flange relative to the horizontal axis, as illustrated in Figure 3.1.

As discussed in Chapter 2, maximum flange angle is directly related to the wheel L/V ratio required for wheel flange climb. A higher flange angle has a lower risk of flange climb derailment. Therefore, it is very important to clearly denote the flange angle in the wheel drawing. In a given manufacturing tolerance range, the flange angle should not be smaller than a specified minimum required value.

3.1.2 Wheel Flange Length

Wheel flange length (Len) is defined as the length of flange starting from the beginning of the maximum flange angle to the point where flange angle reduces to 26.6 degrees (see Figure 3.2).

Also discussed in Chapter 2, wheel flange length is related to the distance limit of flange climb. A longer flange length has a lower risk of flange climb derailment. Therefore, it is equally important to clearly denote the flange length in wheel drawings.

It is also useful to denote the maximum flange-angle length L_0. In Section 2.7 of Appendix B, the two phases of flange climb related to L_0 and Len are discussed. The distance that the wheel can stay in contact with the rail in the section with maximum flange angle during flange climb depends on the length of L_0, wheel L/V ratio, and wheelset AOA.

3.1.3 Wheel Tread Taper

The wheel tread taper, which results from the wheel radius reduction from flange root to tread end (Figure 3.3), provides wheels with a self-centering capability and also helps wheels steering on curves. However, high slope of

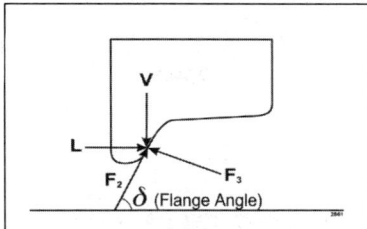

Figure 3.1. Flange angle and flange forces.

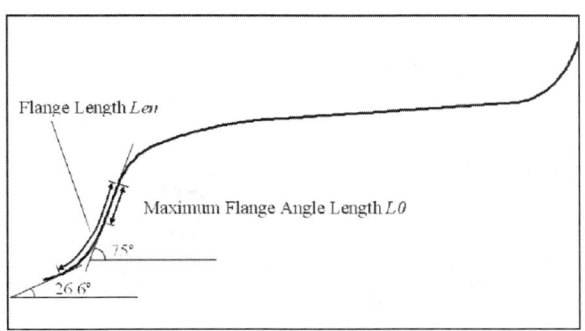

Figure 3.2. Notation of wheel flange length.

taper can increase the risk of vehicle hunting above certain speeds.

When a straight line is designed for the wheel tread, a ratio of radius reduction versus length can be used to denote the taper. For some wheels designed with arcs in the tread section, an equivalent slope may be approximated.

3.1.4 Coordinates of Wheel Profile

The designed wheel profiles are generally described by a series of circular arcs and straight lines. For the convenience of both wheel/rail contact analysis and vehicle modeling, it

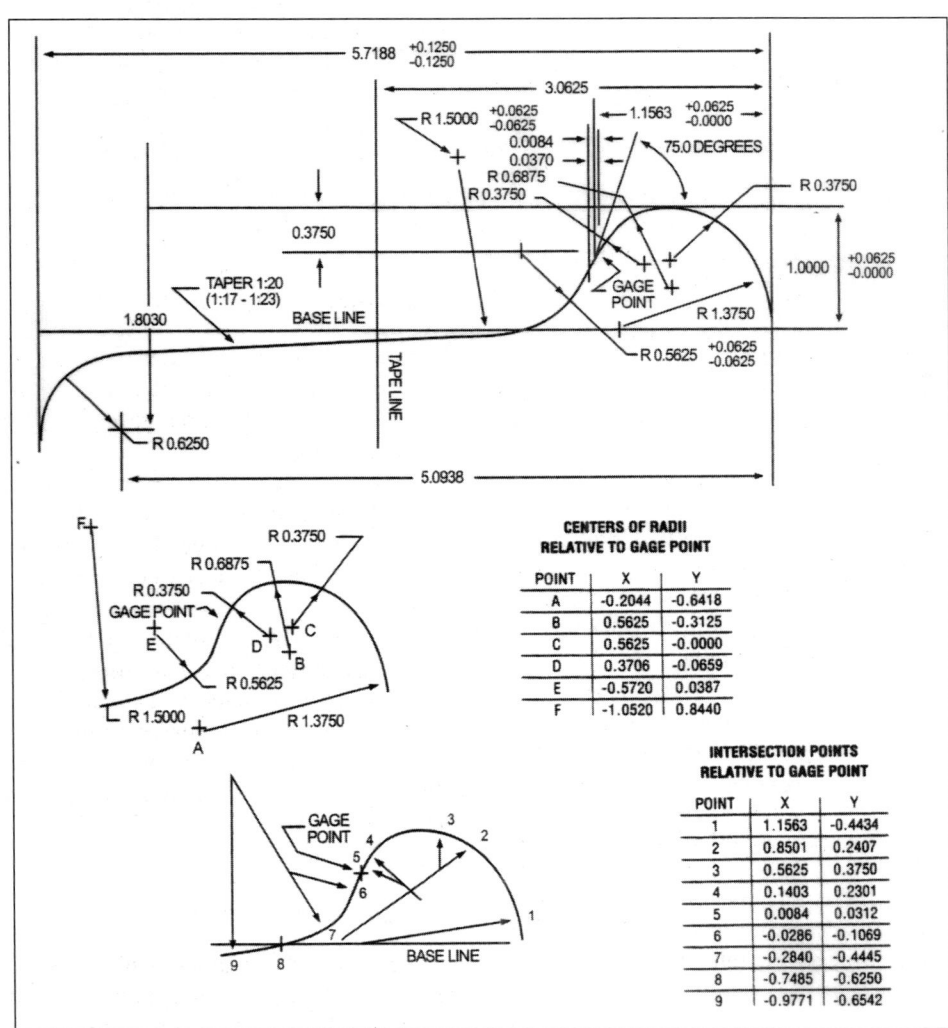

Figure 3.3. AAR-1B wheel profile drawing.

is suggested that the coordinates of intersection points and arc centers be listed on the drawings. Figure 3.3 shows a drawing of the standard AAR-1B wheel profile with two tables listing those coordinates (3).

3.2 WHEEL/RAIL PROFILE MEASUREMENT AND DOCUMENTATION

Measuring wheel and rail profiles is a common means of collecting the information needed for making maintenance decisions. Profile measurement becomes especially important for diagnosing problems due to poor wheel/rail interactions, such as poor vehicle curving, vehicle lateral instability, flange climb derailment, and excessive wheel/rail wear.

However, the measurements are only useful when they are properly taken. Distortion of actual profile shapes can provide wrong information on the cause of problems or the need for maintenance. Good documentation of the measurements can provide a complete and systematic view of the performance of the wheel/rail system in the operation.

3.2.1 Profile Measurement Devices

3.2.1.1 Measurement Gages

The most common devices used in transit operations for wheel measurement are the so-called "go/no-go" gages. These gages are used to measure wheel flange height and flange thickness. A wheel exceeding the limits defined by the gage is either re-profiled or condemned according to a dimension limit, such as wheel rim and flange thickness.

The gages are generally different for the different wheel designs. Figure 3.4 gives an example of a gage from the *Field Manual of AAR—Interchange Rules* (4), used to measure wheel flange thickness. Figure 3.5 shows a gage for measuring the wear on a specific type of rail. The one shown is for rails with the AREMA 136 RE rail profile. Readings of the

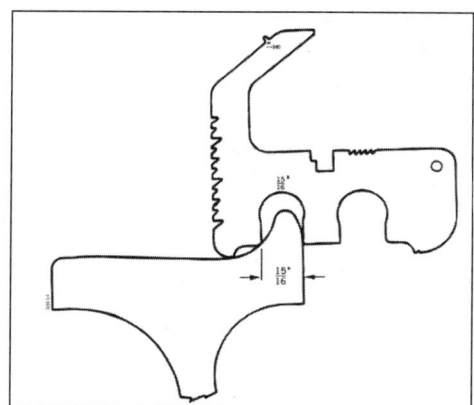

Figure 3.4. Example of gage to measure wheel flange thickness.

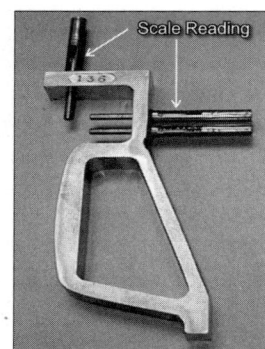

Figure 3.5. Measuring gage for AREMA 136 RE rails.

scale pins on the gage give the wear amount of a rail at three positions.

The above types of gages or similar ones can only provide rough wear information of the measured profiles. For conducting wheel/rail contact analysis, the measurements of complete profiles are required.

3.2.1.2 Profile Contour Measurement

Profile contour measurement gives complete shapes of wheel and rail. For wheels, it can start from the back of the wheel flange to the end of the wheel tread, and for rails, the measurement encompasses the whole shape of the rail head. Figure 3.6 gives examples of wheel and rail profile measurements.

In the past decade, several new profile measurement techniques have been developed. They may be categorized as mechanical, optical, and laser based. Many of them are portable and manually operated. In recent years, automated onboard and wayside measurement systems have been developed. The capability, accuracy, and cost vary for the different types of device.

3.2.2 Effect of Measurement Accuracy on Wheel/Rail Contact Assessment Quality

The accuracy of profile measurement is important to wheel/rail contact assessment or wear analysis, which generally is the purpose of requiring profile measurements. The major factors that may affect the measurement accuracy include the following:

- Calibration
- Setting position of the measurement devices
- Surface cleanness

The calibration of a device sets the measurement accuracy relative to the device origin using a provided template. Each

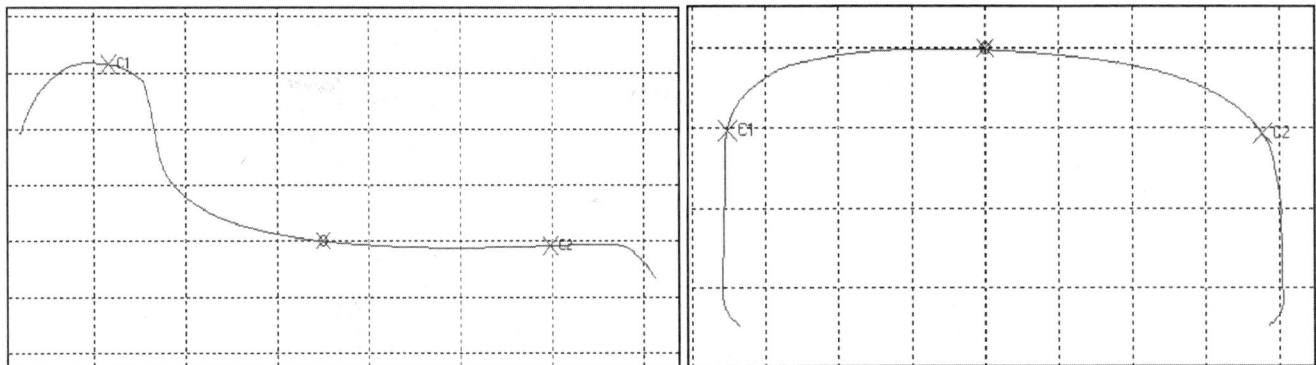

Figure 3.6. Examples of wheel and rail profile contour measurements.

type of measurement device has its own specified calibration procedures. For some devices, each unit has its own calibration file to adjust any error that may be induced by manufacturing tolerance. If the calibration has not been performed properly, the measured profile may be significantly distorted from the real shape. In the example shown in Figure 3.7, two wheel profiles were measured at exactly the same cross section of a wheel using two types of measurement devices. The improperly calibrated device (the lower profile in the figure) measured the wheel profile with a rotation relative to the real shape (the upper profile), which significantly changed the wheel tread slope. When the rotated wheel profile is used to calculate the contact geometry with a rail, the calculated contact situation would also be different from the real condition.

Improper setting of the device can also cause profile distortion. In general, a position plane for wheels, which could be different on various devices, must completely line up with the flange back where there is generally no wear and be perpendicular to the track plane. For rails, it is required that the measurements are relative to the track plane and perpendicular to the longitudinal direction (along track).

Figure 3.8 shows examples of two wrong settings. The left one shows an improper setting of the position plane at the flange back, which causes a rotation of the measured profile. The right one shows where the measurement device is not set perpendicular to the rail, which causes skew of the measured profile. With the skewed condition, the measured profile may be wider than the actual shape. This can distort the actual contact positions and contact radii in the analysis.

Dust, lubricant residue, or other contaminants from the operating environment adhering to the surface can also affect the measurement results. Large pieces of contaminants can distort the measurement shapes, and small pieces can introduce small distortions into the measurements. In profile analysis programs, the measured profiles are commonly transformed from the measured X-Y coordinates into mathematically described shapes. The data variation caused by the debris on measurement surfaces increases the error band of this mathematical transformation.

Therefore, although different profile devices may require different attention, three major procedures for taking profile measurements need to be followed uniformly for portable devices:

- Calibration of measurement devices before taking measurements.
- Cleaning of measurement surface before taking measurements.
- Proper position of measurement devices on wheels or rails to be measured.

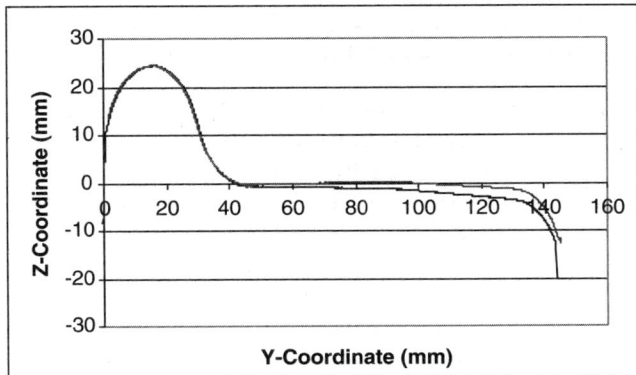

Figure 3.7. Profile distortion caused by improper calibration.

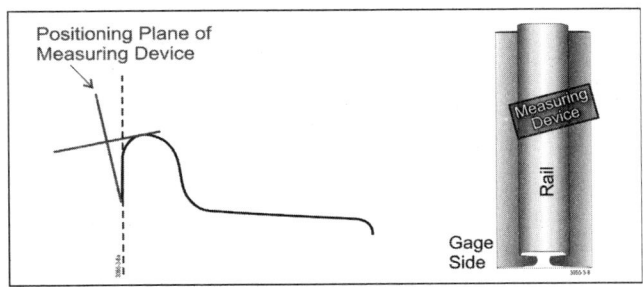

Figure 3.8. Examples of improper setting of measurement devices.

For a portable rail measuring device, a leveling bar is usually used to measure rail gage and hold the measurement device in the correct orientation relative to the track plane. If there is no mechanism for holding the device in the correct orientation, then a direct measurement of the rail cant angle should also be made.

The automated measurement systems generally require more complicated calibrations, as well as additional mathematical smoothing and filtering.

3.2.3 Documentation of Measurement

Good documentation of measured profiles is useful for obtaining a system view of wheel/rail profile conditions, combined with the geometry and contact analysis results. It is especially helpful for tracking profile changes to determine the wear patterns and wear rates, tracking the variations of contact situations to determine the maintenance need, and identifying the performance trends in vehicle types or track sites.

Depending on the purpose for making the profile measurements, other information related to the measurements may also need to be recorded, such as surface conditions (shells, spalls, and head checking), lubrication conditions, tie/fastener conditions at the measurement site for rails, and vehicle condition for wheels. Tables 3.1 and 3.2 give examples of documenting wheel and rail measurements. More columns can be added for additional information.

3.3 WHEEL/RAIL PROFILE ASSESSMENT

How measured wheel and rail profiles are assessed should be based on the objectives of the analysis. They are generally related to the following issues:

- Making maintenance decisions
- Studying wear processes and wear rates
- Studying contact conditions
- Studying wheel/rail interactions

Often, in maintenance decisions, not only the profile shapes are considered but also the surface conditions. In transit operations, flat spots on the wheel surface are one of the common reasons for wheel re-profiling. Corrugation and surface defects due to rolling contact fatigue are among common reasons for rail grinding. In this section, only those assessments related to wheel/rail profiles are discussed.

3.3.1 Dimension Assessment

3.3.1.1 Wheel Flange Height and Flange Thickness

Wheel flange height is defined as the distance from the flange tip to the wheel tread taping line (see Figure 3.9). It is an indicator of tread wear and could also be used as an indicator of rim thickness. Wheel flange thickness is defined as the flange width at a specific height above the taping line. It gives an indicator of flange wear. The minimum flange

TABLE 3.1 Recording example of wheel measurements

Record of Wheel Measurements									
Measurement Date	Measurement Location Shop/Line	File Name of Measurement	Vehicle Number	Axle Number	Left/ Right	Date Last Turned	Mileage Since Last Profiled	Designed Profile	Observations
4/10/04	Shop1/Green	04102004-0010.whl	708932	3	L	2/25/02	50,000	ST1	Surface Shelling

TABLE 3.2 Recording example of rail measurements

Record of Rail Measurements									
Measurement Date	Measurement Location MP/Line	File Name of Measurement	Curvature (degree)	High/ Low	Gage (in.)	Date Last Ground/ Laid	Number of Axle Passes since Last Grinding/Laid	Designed Profile	Observations
4/10/04	18.6/Green	04102004-0010.rai	5	H	56.6	6/22/03	30,000	115 RE	Head checking on rail gage. Poor lubrication

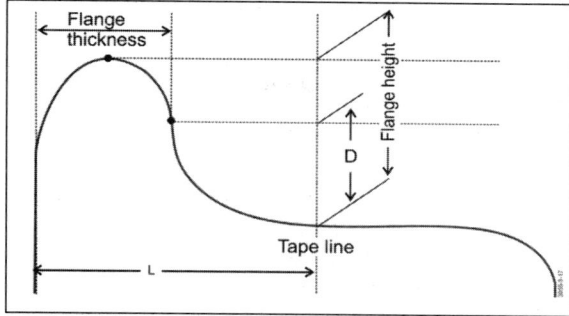

Figure 3.9. Definitions of flange height and flange thickness. (L is the distance from wheel back to the tape line [or datum line], D is the position where the flange thickness is measured.)

thickness limit ensures the bending strength of the flange when subjected to dynamic forces. There are different designs of wheels adopted in transit operations with different dimensions. Each type of wheel should have specifications on the limiting values of flange height and thickness. These specifications should be followed for conducting maintenance.

3.3.1.2 Wheel Tread Hollowness

The wheel tread hollowness is defined by placing a horizontal line at the highest point of the end of the wheel tread. The maximum height from the tread to this line is the value of hollowness (see Figure 3.10). Hollow-worn wheels can have very negative effects on vehicle performance (5, 6). Although rules for removing hollow-worn wheels are still in the process of being established, North American interchange freight service now has a general aim to eventually remove wheels with tread hollowing greater than 3 mm. Transit operations should have a smaller allowed tread hollow limit than freight service not only for operational safety but also for ride quality.

3.3.1.3 Rail Head/Gage Metal Loss

The limit of rail head/gage area loss defines the minimum rail cross sectional area allowed in service. This limit ensures rail has sufficient strength under load and provides adequate guidance for wheels running along the track. The limiting loss of area should be specified based on the vehicle load, track curvature, and track condition.

The head or gage losses measured by the gage in Figure 3.5 are indicated by the graduations on the pins. Using the profile contour measurement device, it is convenient to overlay the worn profile with the new and to compute the area loss, as shown in Figure 3.11.

Most profile contour measurement devices now have software that can quickly process a large group of measured wheels and provide results for wheel flange height, flange thickness, tread hollowing, and other geometry parameters on a spreadsheet.

The rail head material loss computation requires that the measured rail profiles have a correct orientation relative to the new rail template; previous measurements at the same location can be used to confirm the accuracy of the computation.

3.3.2 Wheel/Rail Contact Assessment

Wheel/rail contact assessment is generally performed to study the effects of wheel/rail interaction on vehicle performance or wheel/rail wear. Depending on its objectives, the analysis can be either static or dynamic.

Static analysis only concerns wheel and rail shapes and their relative positions under a specified loading condition without regard to the vehicle or its motion. The results from static analysis are normal contact stress and parameters of the wheel/rail contact constraints. Dynamic assessment is usually performed using vehicle simulation software, which provides detailed information on wheel/rail interaction, including normal forces, tangential forces, creepages, displacements, velocities, accelerations, and other dynamic parameters for wheel and rail contact patches. Contact parameters resulting from dynamic assessment are not only related to wheel/rail shapes and relative positions but are also

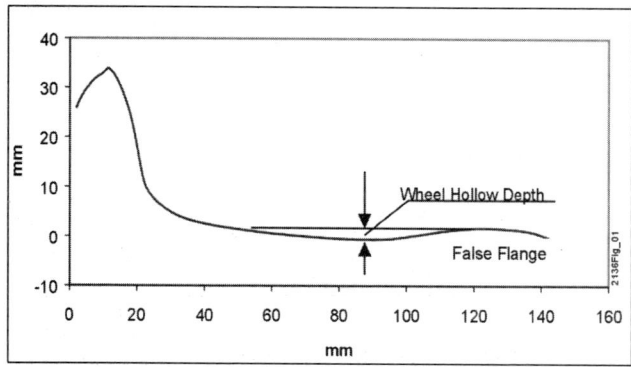

Figure 3.10. Definition of wheel tread hollow.

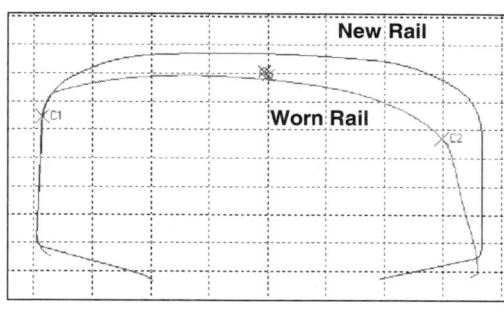

Figure 3.11. Rail head cross sectional area loss.

influenced by speed, car/truck characteristics, and track geometry. The research team has developed a static analysis software program (6) and a dynamic analysis program (7).

The static analysis software can analyze contact situations of many wheelsets against a measured pair of rails or many rails against a measured pair of wheels. This method provides a comprehensive view of wheel/rail contact at a system level. For example, thousands of wheels with different profiles (due to different levels of wear or resulting from different truck performance) could contact a section of rail at different positions and, therefore, could produce different contact patterns and different levels of contact stress. The performance of the majority of wheel/rail pairs is therefore the focus of the assessment.

The distribution of contact parameters can be used to predict likely vehicle performance, wheel/rail wear, and contact fatigue. For example, consider a group of measured wheels contacting a pair of rails measured on a curve. If the rails are judged to have unsuitable profiles due to resulting high contact stress and undesirable contact patterns, then appropriate action can be taken. If only a small number of wheels give unwanted wheel/rail interaction, then it might be best to remove those wheels from service. Alternatively, if many wheels cause problems, then it might be best to re-profile the rail by grinding.

Dynamic assessment is generally performed to study the wheel/rail interaction for specific vehicle/track conditions. Therefore, using wheels on the vehicles being studied would more accurately predict their performance. The contact tangential forces and creepages produced from dynamic simulation can provide more detailed information for the analysis of wear and rolling contact fatigue. A large number of simulations would need to be conducted if a detailed analysis of a large group of wheel profiles was required, such as was needed for the derailment study performed for this report.

In summary, the analysis of a large number of profiles is useful for wheel/rail system monitoring and evaluation. A static analysis generally can produce the required results quickly. Dynamic simulation can provide more detailed information related to wheel/rail interaction under specific conditions. The method that should be selected for the wheel/rail profile analysis depends on the assessment objectives.

The parameters produced from the wheel/rail profile analysis are described in detail below.

3.3.2.1 Maximum Contact Angle

The maximum wheel/rail contact angle depends on the maximum wheel flange angle and the maximum angle of the rail gage face. A wheel profile with a higher flange angle can reduce the risk of flange climb derailment and can have much better compatibility with any new design of vehicle/truck that may be introduced in the future compared to wheels with lower flange angles. Also, with a higher L/V ratio limit (according to the Nadal flange climb criterion), high flange angles will tolerate greater levels of unexpected track irregularity.

Figure 3.12 shows two examples of undesirable relationships between wheel flange angles and the preferred relationship. If rails are worn into a lower gage angle than that of the wheel flange angle or if newly designed wheels have a higher flange angle than existing wheels, a point contact would occur on the wheel flange, and this would result in a maximum wheel/rail contact angle less than the maximum wheel flange angle. The contact situation is likely to be as shown in the left illustration of Figure 3.12 as wheel flanging. If the wheel flange angle is lower than the rail gage angle, the contact situation is likely to be as shown in the middle illustration of Figure 3.12. The right illustration shows the desirable flanging condition where wheel flange and rail gage face wear to similar angles.

3.3.2.2 Contact Positions

Wheel and rail contact have a direct effect on vehicle performance and wheel/rail wear. Contact positions are closely related to wheel/rail profile shapes and influenced by vehicle and track condition. The three typical contact conditions shown in Figure 3.12 are likely to produce different curving forces and rolling resistances. Distribution of contact positions on a pair of rails from contacting a population of wheels gives indications of the likely performance trend.

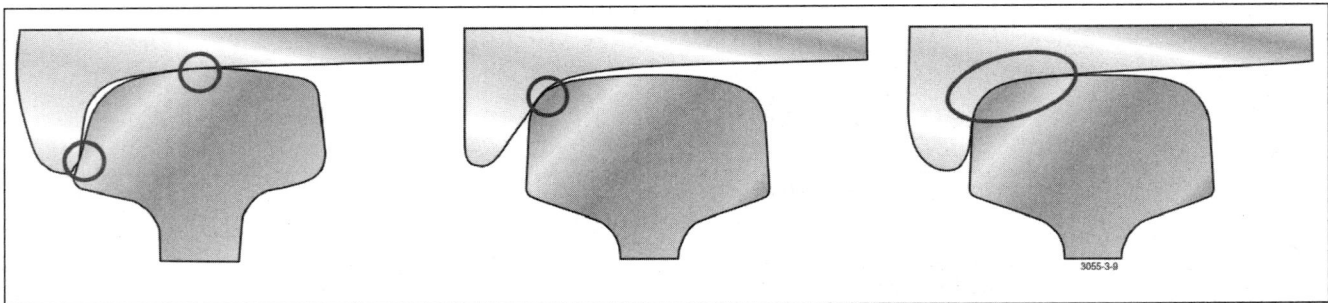

Figure 3.12. Three types of contact related to wheel flange/rail gage angles.

Figure 3.13 shows an output example from the static analysis software with 112 wheelsets, which were measured on trains that had passed over a pair of rails measured on a 7 degree curve. A wheelset lateral shift between 0.3 and 0.5 in. was assumed for the computation. That is the lateral shift range for wheel flanging on this degree of curve. The dots in the figure show the distribution of contact positions on the rails from those 112 axles and the level of contact stress. The high rail showed a trend of conformal contact indicated by the relatively even distribution of dots and the number of wheels contacted at each band. While wheels were flanging on the high rail, the low rail showed highly concentrated contacts toward the field side. Of the 112 wheelsets, 87 contacted at a distance only about 0.5 in. from the field side and produced high contact stress. In this situation, rail grinding was suggested to correct the low rail shape. Removing metal at the field side of rail can shift contact positions to the rail crown region and reduce contact stress.

By varying the wheel lateral shift range, the distribution of contact positions of leading and trailing axles can be separately investigated, as well as the distributions on different degrees of curves.

3.3.2.3 Contact-Stress Level

Contact-stress level is one of many factors that affect rolling contact fatigue and wear at contact surfaces. Combined with the distribution of contact positions, the distribution of contact stress provides an indication of likely wear patterns and the risk of rolling contact fatigue.

Good wheel/rail profile designs should produce lower contact stress and less locally concentrated contact. Although there are arguments about the critical level of contact stress, the generally accepted level is in the range of 220 to 290 ksi in the rail crown area, and about 480 ksi in the gage face area when considering the effect of lubrication and strain hardening for commonly used rail steels.

In Figure 3.13, the low rail experienced contact stress toward the field side that was much higher than the criterion and should be corrected. The high rail experienced acceptable contact stress toward the gage side, but high stress at the gage corner because of the very small contact radius. High contact stresses in this area combined with the tangential forces can cause the metal to either wear off before microcracks develop to a size that causes concern or else form head checking rolling contact fatigue (RCF) or other defects. Whether wear or RCF occurs depends on the lubrication conditions, tangential forces at the contact patch, and the hardness of the wheel and rail steels.

3.3.2.4 Effective Conicity on Tangent Track

Lateral instability is more likely to occur when there is high wheelset conicity (the ratio of RRD between the left and right wheel over the wheelset lateral displacement). In this circumstance, as speed is increased, the lateral movement of the wheelsets, as well as the associated bogie and carbody motion, can cause oscillations with a large amplitude and a well-defined wavelength. The lateral movements are limited only by the contact of wheel flanges with rail. The high lateral force induced from hunting may cause wheel flange climbing, gage widening, rail rollover, track panel shift, or combinations of these. Vehicle lateral stability on tangent track is especially important for high speed transit operation. With a properly designed vehicle suspension and the modest maximum speeds of most transit operations, high wheelset conicity should not cause vehicle instability, although it can occur.

The effective conicity of wheel/rail contact has considerable influence on the vehicle hunting speed. As wheelset conicity increases, the onset critical speed of hunting decreases. The effective conicity is defined by Equation 3.1 (9):

$$Effective\ Conicity = \frac{RRD}{2y} \quad (3.1)$$

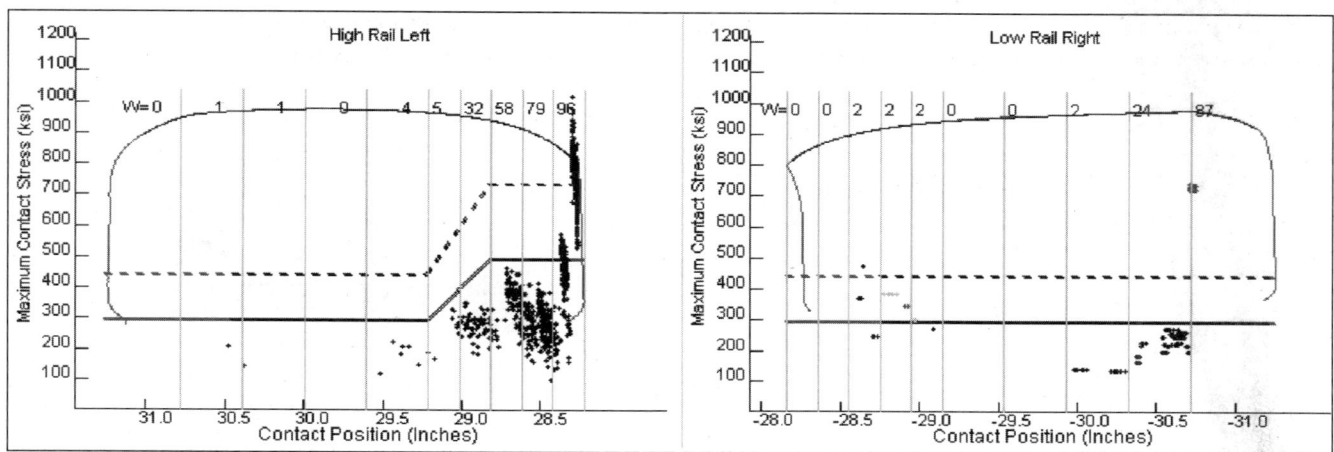

Figure 3.13. Distribution of contact positions and contact stress.

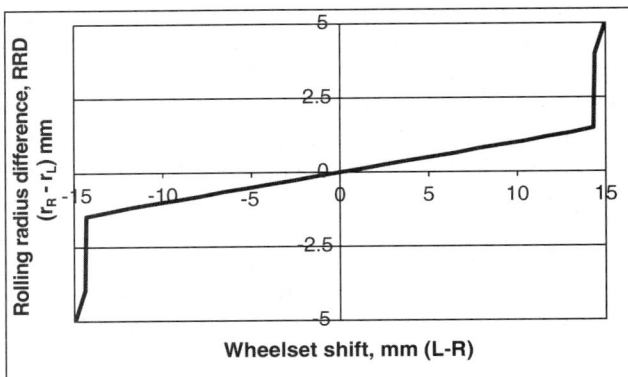
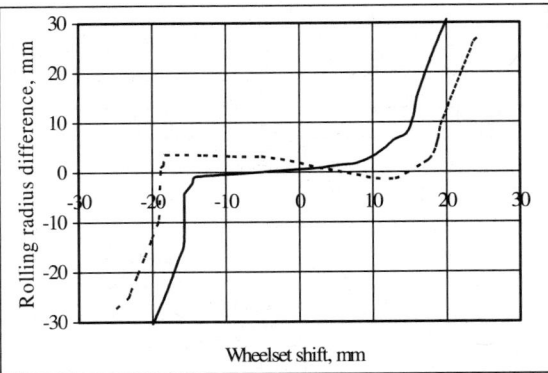

Figure 3.14. RRD versus lateral shift.

where *y* is wheelset lateral shift relative to rail. The left diagram of Figure 3.14 shows an example of RRD versus lateral shift for new wheel contacting new rail. The slope of the straight section before reaching flange is used to compute the effective conicity, which is usually a constant. The right diagram of Figure 3.14 shows two examples of worn wheels contacting worn rails. Under worn wheel/rail conditions, the effective conicity is no longer a constant. Equation 3.1 should be used for each specified wheel lateral shift value and corresponding RRD.

The critical hunting speed is highly dependent on the vehicle suspension characteristics and the effective conicity of the wheel/rail profiles. The maximum conicity that can be tolerated is critically dependent on the vehicle suspension design. As discussed, large wheelset RRD (which can be obtained with high effective conicity) is beneficial to truck curving ability. In comparison, high effective conicity can cause lateral instability in a poor vehicle suspension design, thus limiting maximum operating speed. The wheelset effective conicity should be carefully selected along with the vehicle suspension design to give the optimum compromise between lateral stability and curving performance for each transit system. Although the critical value can be varied by vehicle types, generally the effective conicity should be no higher than 0.3. Note however that RRD and wheelset conicity has no effect on the hunting speed of trucks equipped with independently rotating wheels.

Dynamic analysis and track tests are especially important in introducing new vehicles and/or new profiles into a system to ensure that, for a specific vehicle/track system, the critical hunting speed is above the operating speed.

3.3.2.5 RRD for Curving

For a wheelset with a rigid axle to properly negotiate a curve, the wheel contacting the outer rail requires a larger rolling radius than the wheel contacting the inner rail. The difference in rolling radius between the two wheels of a wheelset is defined as the RRD, as illustrated in Figure 3.15. Without adequate RRD, wheelsets can experience higher AOAs and higher lateral forces (before reaching saturation). As a result, both wheel and rail can experience higher rates of wear. Note that for trucks with independently rotating wheels, RRD has no effect on vehicle curving.

Equation 3.2 computes the required RRD (Δr_s) between two wheels in a solid axle under a pure rolling condition.

$$\Delta r_s = r_0 \left[\frac{R+a}{R-a} - 1 \right] \quad (3.2)$$

where

r_0 = the nominal wheel radius,
R = the curve radius measured to the track center, and
a = half the lateral spacing of the two rails.

Figure 3.16 gives examples of the required RRD under a pure rolling condition (a wheelset without constraints) for three different wheels. The values are related to track curve radius, wheel diameter, and track gage. A gage of 56.5 in. was used in this calculation. Note that the curve radius has been converted to curvature in degrees in this figure.

Equation 3.2 and Figure 3.16 show that a large rolling radius (high effective conicity) provides for improved vehicle steering and reduced wheel/rail wear. In Section 3.3.2.4,

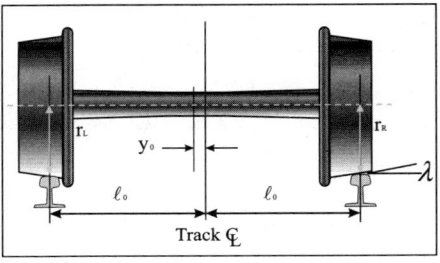

Figure 3.15. RRD (r_l versus r_R).

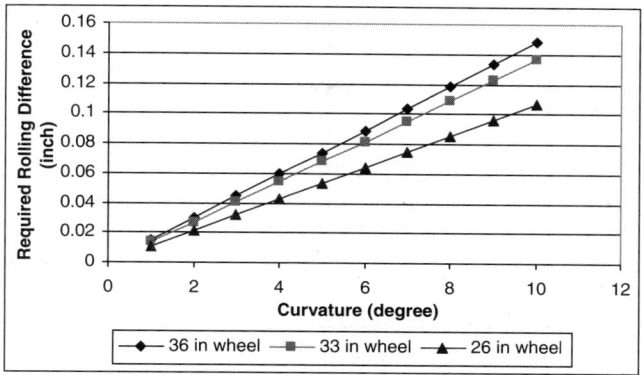

Figure 3.16. Required RRD for pure rolling.

it was shown that a low effective conicity could reduce the tendency for a wheelset to hunt. However, the hunting of a vehicle is also critically affected by the vehicle primary longitudinal and lateral suspension stiffness. Most transit systems operating in North America have relative high primary suspension stiffness, which reduces the tendency to hunt.

For the majority of the time, many transit systems operate at relatively low speeds (below 50 mph) and have many curves. Therefore, curving and consequent wheel and rail wear is likely to be more important than vehicle hunting. RRD and wheel/rail conicity should be optimized for each system based on the suspension parameters for the particular vehicles on each system, standard operating speeds, and the mix of straight and curved track for the system. Different rail profiles can be designed for curved and straight track and the wheel profile designed to optimize performance with those profiles. Analyses of curving and hunting performance using vehicle dynamic computer models is recommended.

RRD on large radius curves (low degrees of curvature). For curves with a radius larger than 2,000 ft (close to 3 degrees of curvature), there is not likely to be hard wheel flange contact. The RRD is mainly dependent on the slope of the wheel tread and the flange throat region before flanging. Figure 3.17 illustrates the rolling radius varying with the tread taper. When the wheel is worn, the rolling radius would not be linearly varying with the wheelset lateral shift.

Take an example of a 33-in. diameter wheelset with a 1:40 taper (0.025 conicity). By assuming that a 0.3-in. wheelset lateral shift relative to the track will nearly cause wheel flanging for this wheel and that the 1:40 taper is maintained in this lateral shift range, an RRD of 0.015 in. will be obtained. Figure 3.16 shows that this level of RRD will achieve pure rolling on a 1-degree curve. If the wheel has a 1:20 taper (0.05 conicity) for the same lateral shift, pure rolling can be achieved on a 2-degree curve with an RRD of 0.03 in. On large radius curves, free rolling generally can be achieved with adequate RRD.

Note that the RRD only from the wheel taper is limited by the lateral clearance allowed between the wheel and rail (which limits the lateral shift). When the clearance is used up, the RRD depends on the shape of the wheel flange throat or flange. The rail shape can also influence the RRD. In Figure 3.18 (shown in an exaggerated way), the low rail B would produce bigger RRD than rail A by taking advantage of wheel taper, assuming the high rail is maintained in the area close to the wheel throat.

RRD on small radius curves (higher degrees of curvature). On curves with a radius smaller than 2,000 ft, wheels on the high rail are likely to be in flange contact. Depending on the wheel/rail profiles, the contact on the outer wheel/rail can be one-point, two-point, or conformal, as illustrated in Figure 3.12. The rolling radius at the wheel flange root (or slightly down the flange) can be 0.2 to 0.5 in. larger than that on the wheel tread depending on the flange height and wheel shape. For example, according to Figure 3.16, a 0.3-in. RRD can provide free rolling on curves of about 20 degrees (with a curve radius of about 300 ft and a gage of 56.5 in.).

Again, the clearance between wheel and rail also limits the maximum RRD that can be reached due to limited lateral shift allowed. For example, consider a railroad that only allows 0.08 in. (2 mm) of wheel and rail clearance. The RRD in this situation would be considerably smaller because both wheels are possibly contacting the rails in the flange throat and on the flange faces.

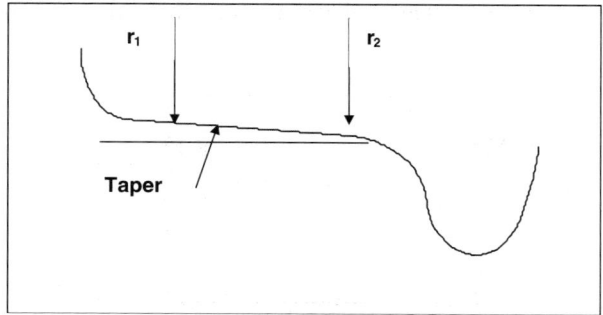

Figure 3.17. Rolling radius varying with wheel tread taper.

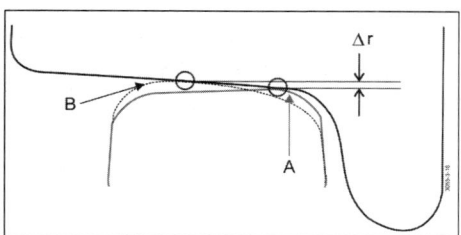

Figure 3.18. RRD affected by low rail shapes. (Δr is the radius difference caused by rail contacting a wheel at different positions.)

However, on small radius curves, free rolling generally cannot be achieved. The major influences come from truck primary yaw stiffness and clearance between truck frame and axle bearing adapters. The wheel/rail lubrication condition can also influence the possibility of free rolling. Therefore, the RRD to avoid flange contact as computed by Equation 3.2 can only be considered as the base requirement from wheel and rail profiles.

In transit operations, especially in urban areas, some curves have very tight radii. As a result, it is not possible to achieve the required RRD from the wheel and rail geometries. Wheel sliding and higher wear rates become common in those sections. A softer primary suspension and lubrication may improve the situation.

In curving, if there is only one-point contact on the outer wheels, the contact RRD is relatively easy to determine. However, if there is two-point contact, especially where this condition is severe on the outer wheel, the evaluation of vehicle curving ability from the view of wheel/rail profiles is more complicated. This condition is discussed in the next section.

It can be seen from the above discussion that requirements of RRD for curving and lateral stability are conflicting. Proper curving requires higher RRD, which results from higher effective conicity, and lateral stability requires lower effective conicity. The required compromise has to be achieved by adequately designed wheel profile and ground rail profiles. Note that wheels run over all sections of rail in a specified system while rail is locally stationed. Therefore, adjusting rail profiles based on the local operational emphasis can improve both curving and lateral stability. This issue will be further discussed in the section of ground rail design.

3.3.2.6 Effects of Two-Point Contact

Two-point contact is defined as a wheel contacting the rail at two clearly separated locations. Severe two-point contact usually has one contact point on the wheel tread and the other on the flange. As discussed in Section 4.5.1 of Appendix A, severe two-point contact is not desirable in curving since it reduces the wheelset's steering ability because the longitudinal creep forces generated at these two points can act in opposite directions.

The size of the gap between wheel and rail during flanging, d, can be used as an indicator of the severity of two-point contact. The larger this gap, the more severe will be the two-point contact (that is, the two contact points will be farther apart and the wear-in period will be longer). The National Research Council, Canada, defines the nature of the contact according to the size of the gap:

- If d is 0.1 mm or less—close conformal contact,
- If d is 0.1 mm to 0.4 mm—conformal contact, and
- If d is 0.4 mm or larger—nonconformal contact.

Conformal and close conformal contacts are desirable on curves for producing lower lateral forces and rolling resistance. Nonconformal contact is shown in Figure 3.19, d being 1.2 mm, which is severe two-point contact.

In a system, if severe two-point contact is the trend of wheel/rail contact in curving, the wheel- and rail-wear rate is likely to be high due to high creepages and creep forces at the contact surfaces. In sharp curves, the risk of rolling contact fatigue could also be higher than that of a conformal contact situation.

Although severe two-point contact is to be avoided, too much conformality, such as that which occurs with very worn wheels and rails, can also have drawbacks. Wide bands of conformal contact between the wheel and rail in the region of the gage shoulder have been implicated as a potential contributor to RCF (rail gage corner cracking), especially in shallow curves where the wheels are not running in flange contact (10). Current hypotheses suggest that this occurs for vehicles with relatively stiff primary suspension in both lateral and longitudinal directions.

Although research is ongoing in this area, potential methods for controlling this form of RCF may include the following:

- Optimizing wheel/rail profiles to improve vehicle steering by
  - Reducing the width of the contact band in the rail gage shoulder or
  - Increasing the wheel conicity in the flange root area, which gives a smoother transition of contact from rail head to the gage shoulder;
- Optimizing vehicle suspension stiffness to improve vehicle steering;
- Applying friction modifiers and/or lubricants to the rail head to reduce wheel rail forces; and
- Using harder rail steels.

Hence, compatible wheel and rail profiles are critical for a system to reach desirable contact patterns. Figure 3.20 gives an example of gap distribution for a group of measured wheels contacting a pair of measured worn rails in the same system. Conformal contact was reached for these combinations for the majority of values below 0.4 mm.

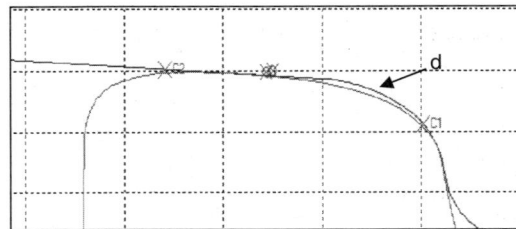

Figure 3.19. Illustration of gap between wheel flange root and rail gage.

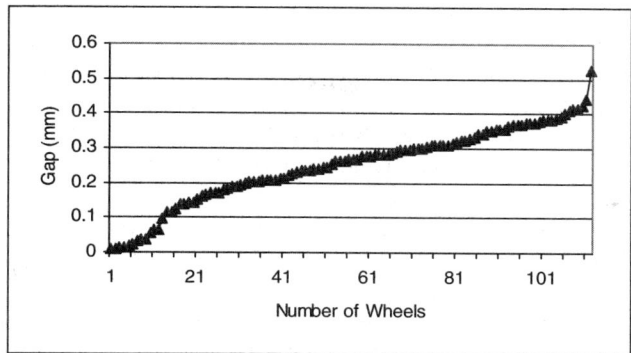

Figure 3.20. Example of distribution of contact conformality.

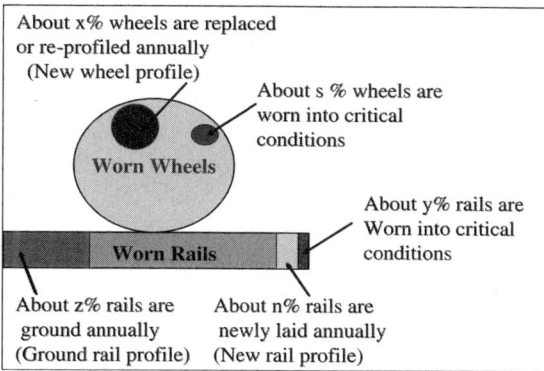

Figure 3.21. Illustration of distribution of wheel and rail conditions in a system.

3.4 UNDERSTANDING IMPORTANT STAGES OF WHEEL/RAIL CONTACT IN A SYSTEM

As listed in Table 3.3, the wheel/rail contact situations in a system can generally be categorized into several important stages. Those stages usually exist in parallel in a system due to different life and wear levels of wheels and rails, different loads or capacities between lines, and different maintenance processes.

Appreciating the conditions of these important stages of wheel/rail contact in a system can provide insight into the improvement of wheel/rail interaction and can assist in the management of wheel/rail maintenance. Figure 3.21 illustrates distribution of wheels and rails assumed in a system.

Desirably, the dominant contact condition in a system should be stable, worn wheels contacting stable, worn rails. Starting with compatible new wheel and rail profiles, contact of stable, worn wheels and rails should produce desirable contact features and should last a relatively long period without other disturbances.

The contact conditions, listed in Table 3.3, are further discussed in the following sections to emphasize their distinguishing features and the attention that may be required.

3.4.1 Initial Contact Conditions—New Wheel Contacting New Rail

Every year in a system, new wheels will replace condemned wheels, and some sections of rail may be re-laid,

TABLE 3.3 Important stages of wheel/rail contact

Important Stages	Related Wheeland Rail Profiles
Initial contact conditions	New wheels contact new rails
Stable contact conditions	Stable worn shapes of wheel and rail
Contact conditions of new or newly trued wheels	New wheels contact rails from new to worn
Contact conditions after rail grinding	Wheels from new to worn contact ground rails
Critical contact conditions	Wheel/rail shapes indicate risk of derailment and cause significant damage to the system

which will lead to the condition of a new wheel contacting a new rail. Of course, it is also the contact condition of a newly opened line.

The initial condition determines the likely wear patterns of wheels and rails, the wear-in period, and the effects of wheel/rail profiles on vehicle performance. The initial condition should be carefully considered and analyzed; especially for a new rail system. All contact parameters discussed in Section 3.3 should be assessed and documented. Simulation and track test should also be performed to ensure that the new wheels and rails provide desirable dynamic performance under specified vehicles and tracks.

Some transit systems have wheel/rail profile standards that were established many years ago. Awareness of the initial contact conditions of those profiles would contribute to an understanding of what can be expected in wheel/rail interaction and wear and what improvements in profiles can enhance wheel/rail interaction.

Any new wheel and rail profile combinations starting with severe two-point contact will produce higher wear rate, longer wear-in period, and poorer curving in the initial stage. Possibly the new combinations provide better lateral stability. Section 4.5.1 of Appendix A gives examples of three types of initial contact conditions in surveyed transit systems.

3.4.2 Stable Contact Conditions—Stable Worn Shapes of Wheel and Rail

Stable contact is considered to be the desirable equilibrium condition. When this stage is reached after the wear-in period, wear rate and contact stress should be relatively low due to a conformal contact situation at both wheel tread/rail crown and wheel flange throat/rail gage areas. Without disturbances from sudden changes on vehicles and tracks (such as changes in vehicle yaw stiffness due to damaged dampers or rail cant changing due to tie plate cutting), the stable condition should continue for a reasonably long period.

Note that different initial contact conditions may lead to different equilibrium situations. These conditions likely inherit problems from the initial contact, such as low flange angle.

3.4.3 Contact Conditions of New or Newly Trued Wheels in Worn Rails

Transit operations have to true wheels (return them to the shape of a new wheel) somewhat more often than freight service due to wheel flats. Wheel flats can be caused by frequent braking and acceleration or by wheel sliding due to contaminated track (see Appendix A, Section 4.7).

The equilibrium of stable wheel/rail contact is lost once new wheel profiles are introduced. New wheels need a wear-in period to reach the equilibrium state with existing rail profiles. During this period, the vehicle curving performance is likely to be poorer than during the stable stage because of the likelihood of two-point contact conditions. The vehicle lateral stability is likely to be better than at the stable stage due to lower effective conicity.

3.4.4 Contact Conditions after Rail Grinding

Rail grinding is often conducted in transit operations to remove rail corrugations and surface defects and sometimes to improve wheel/rail contact. Like wheel truing, rail grinding may also change the equilibrium contact conditions. Rails are usually not ground back to the new rail shape. Therefore, the contact condition after rail grinding is influenced by the designed ground rail shapes and the accuracy of rail grinding. After rail grinding, the wheel/rail contact could be completely different from the previous three conditions. Sometimes, the contact condition could be even worse than that before grinding due to improperly ground rail template(s) or poor grinding accuracy.

Assessment of the contact conditions of grinding templates (designed ground rail profiles) and rail profiles after grinding should be done using the representative wheels that run past the grinding sections. This will ensure that the grinding templates are adequate for the grinding sections and the shapes of templates have been closely reproduced.

3.4.5 Critical Contact Conditions and Associated Wheel/Rail Profiles

Critical contact conditions are defined as wheel/rail profiles that may cause significant damage of wheels and rails or considerably increase the risk of derailment. The associated wheel and rail profiles may include these conditions:

- Thin flange
- Low wheel flange angles
- Hollow wheels
- Low rail gage angles
- Low rail with field side contact
- Significant loss of rail cross section

When wheels and rails wear into the critical shape, they should be either re-profiled or replaced.

3.5 WHEEL RE-PROFILING

Wheel truing is a process for re-profiling the wheel shape and removing surface defects like flats, spalls, and shelling. Two types of wheel truing machines are commonly used. The milling type has a cutting head with many small cutters. The arrangement of the cutters forms the wheel profile. The lathe type has a wheel profile template. The single cutter cuts the wheel by following the shape of a template.

Three major aspects require special attention in wheel truing: tolerances, surface finishing roughness, and lubrication after truing. Here, it is assumed that the profile accuracy of the cutting tools or template has been reached since they are usually professionally preset.

3.5.1 Tolerance between Wheels, Axles, and Trucks

In the transit systems involved in this survey, the wheel diameter truing tolerances ranged from 1/16 to 1/8 in. for wheels on an axle, 1/4 to 1 in. for axles within a truck, and 1/4 to 1 in. for trucks within a car.

In general, the manufacturer's specification on wheel diameter differences for axles within a truck and for trucks within a car should be followed for both powered and unpowered axles.

The difference in diameter for wheels on a (coupled) axle could either lead to the truck running off-center if two axles within a truck have similar patterns of diameter difference or cause the truck to rotate or yaw if the two axles within a truck have different patterns of diameter difference.

The wheel diameter difference from truck to truck within a car may affect the load sharing patterns at the truck center pivot and produce different wheel-wear patterns, but only if the diameter difference is significant. Smaller tolerances would provide a better defined vehicle running behavior. The diameter difference for wheels in a coupled axle is most critical for truck performance. Some European railway systems only allow a 0.02 in. (0.5 mm) difference in diameter for wheels within a coupled axle. Considering the capacity of wheel truing machines currently available in some transit systems, this tolerance should not exceed 1/16 in. (1.5875 mm) difference in wheel diameter. The diameter difference for axles within a truck is critical for the powered axles. Under the same axle rotating speed, both axles may slide due to the wheel diameter differences. This is especially true for mono-motor trucks, because mechanical coupling between axles force the axles to rotate at the same speed. Therefore, the truck manufacturer's specified tolerances on the wheel diameter for powered cars should be strictly followed. The profiling tolerances should not be difficult to achieve if the truing machines are properly maintained.

3.5.2 Surface Finish Requirements

Several systems have reported flange climb derailments occurring at curves or switches in yards just after the wheels had been trued. This type of derailment may have been a result of the required maximum flange angle not being obtained, but was more likely caused by excessive wheel surface roughness after wheel truing. Figure 3.22 shows examples of wheel surfaces just after truing and after many miles of running.

Generally, the coefficient of friction for dry and smooth steel-to-steel contact is about 0.5. The effective coefficient for a rough surface could be much higher. For example, if the coefficient reaches 1.0, the L/V limit (Nadal criterion) would be 0.5 for a 75-degree flange angle and 0.3 for a 63-degree flange angle. Therefore, the rough surface produced by wheel truing could significantly reduce the L/V limit for flange climb. The low flange angle further increases the derailment risk.

Several remedies may improve the surface condition:

- Frequently inspect the cutting tools, especially for the milling type machine. Dulled tools can produce a very rough surface. Sometimes, the grooves are obvious.
- Address the final surface turning. In this step, there is no significant material removal but a light cut is used for smoothing the surface.

3.5.3 Lubrication after Profiling

Lubrication after truing can also be an effective way to prevent flange climb derailment with newly trued wheels. Reducing the friction coefficient at the wheel/rail interface can increase the L/V limit for flange climb. The sharp asperities on the wheel surface after truing may quickly deform or wear off in operation due to very high locally concentrated contact stresses. After operating for some time, the wheel surface should be in a smoother condition. Light lubrication can help wheels safely pass through this rough to smooth transition.

Lubrication can be performed as one of the procedures of wheel truing or applied using wayside lubricators installed on the curve in yards. Other techniques, such as onboard lubrication systems, can also be employed.

3.6 WHEEL PROFILE DESIGN

Given the effect of wheel profiles on vehicle performance and wear discussed above, the requirements for wheel profile design are clear and can be summarized as follows:

- The design must meet the dimension requirements for a specific system.
- Higher flange angle is necessary to reduce the risk of flange climb derailment on curves.
- Effective conicity must be selected to give the optimum compromise between curving and stability requirements for the particular vehicle design and transit system.
- Severe two-point contact with rail to improve curving and reduce wear should be avoided.
- High contact stress should be avoided.

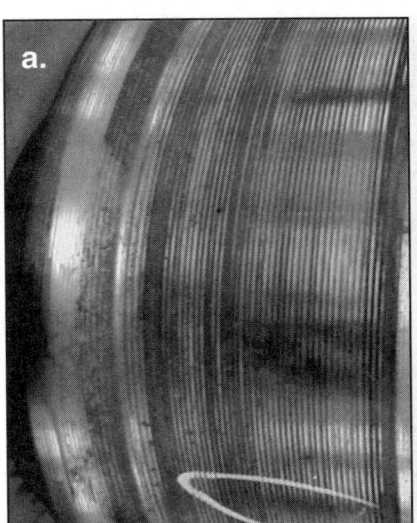
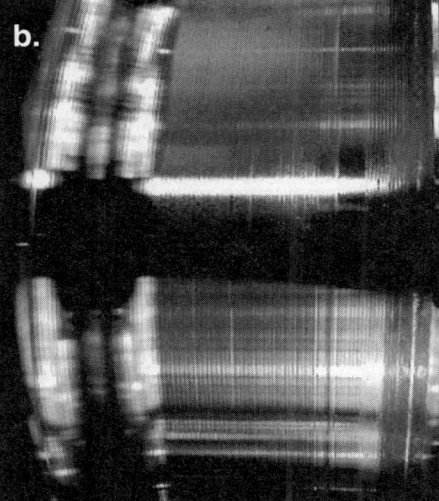

Figure 3.22. Comparison of wheel surface roughness. ([a] surface after wheel re-profiling from milling type machine, [b] surface after wheel re-profiling from lathe type machine, and [c] surface of wheel back from operation with a flat spot.)

Note that the designed wheel flange angle currently used in North American transit operation ranges from 60 to 75 degrees. *TCRP Report 57: Track Design Handbook for Light Rail Transit* (*11*) proposed a wheel flange of 70 degrees based on Heumann's design. The APTA *Passenger Rail Safety Standard Task Force Technical Bulletin* (*12*) provided guidance on reducing the probability of wheel-climb derailment, suggesting a minimum wheel flange angle of 72 degrees (suggested tolerances are +3.0 degrees and −2.0 degrees).

A new wheel profile may be requested for a completely new rail system starting with new wheels and new rails. The design emphasis under this condition is to establish desirable starting wheel/rail contact features to help new vehicles meet their performance requirements. Simulations and trial tests should be conducted on vehicles with the new wheel profiles under the specified operating conditions to ensure that the specified requirements have been meet. The likely wear patterns may be predicted to further determine the vehicle performance under worn wheel/rail shapes.

A new wheel profile may also be requested for an existing rail system with worn wheels and rails. Under this condition, the existing worn wheel/rail shapes should be taken into consideration when designing the new wheel profile, for example when adopting a new wheel profile with a higher flange angle to replace the existing low-flange-angle wheel. If the profile change is significant compared to the existing design, it is likely that an interim profile (more than one, when necessary) will be needed to gradually approach the desired profile. Further, a transition program should be carefully planned by considering the capacity of both wheel truing and rail grinding on the system. An example is discussed in Section 4.2.3 of Appendix A.

3.7 GROUND RAIL PROFILE

In general, North American transit operations use American Railway Engineering and Maintenance of Way Association (AREMA) standard rails (*13*). The AREMA 115-pound (115RE) and the 132-pound rail (132RE) are two types of rail that are commonly used on transit lines, especially for newer systems. Some old systems have lighter weight rails (90- to 110-pound rail) still in use on their lines.

Rail profiles are generally not ground back to the new rail shapes. The ground rail shapes determine the contact condition after rail grinding or determine the variations of contact condition compared to the situation prior to grinding. Therefore, properly designed ground rail profiles are critical for producing minimum disturbances to a profile-compatible system. This would result in a short wear-in period to reach the equilibrium stage.

Sometimes, rail grinding is conducted for a major correction of rail shapes, such as increasing contact angle or relocating the contact band. Under these conditions, the ground rail design must take the wheel shapes in the system into consideration to ensure that the objectives will be achieved.

Ground rail profiles can be designed differently for straight and curved tracks. This is likely to be the case for high speed operations because of a strong emphasis on lateral stability on straight track. On straight track, the goal is mainly to achieve low effective conicity, thereby raising the speed at which vehicle hunting could start. In curved track, the main emphasis may be placed on improving vehicle steering and reducing lateral forces and rolling resistance.

3.7.1 Ground Rail Profile for Straight Track

In straight track, high values of effective conicity can lead to hunting at lower speeds for lightweight vehicles. Hence, one goal of profile design should be that a general low effective conicity trend is maintained for a large population of passing wheels that may have varying tread slopes due to different levels of wear.

One way to lower conicity (actually, to lower the RRD of two wheels) from the ground rail is to reduce the contact at the wheel flange throat by producing a strong, two-point condition when the wheel flanges, as illustrated in Figure 3.23. Then, the rail has no chance to contact the flange throat, thus reducing the variation of rolling radius.

Care needs to be taken when grinding the rails in straight track to avoid concentrating contact in just one portion of the wheel tread. Concentrated contact can lead to excessive wheel hollowing.

3.7.2 Ground High Rail Profile

The ground high rail shapes generally should be close to the stable, worn high rail shapes at the rail gage shoulder and corner. Severe two-point contact should be avoided because it produces poor steering.

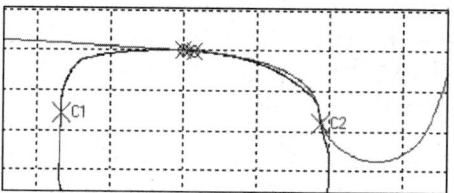

Figure 3.23. Example of reducing effective conicity by controlling contact position.

If grinding were only conducted on the stable, worn rail shapes to remove corrugations or surface defects, maintaining these stable shapes would lead to a minimum disturbance to the existing conformal contact. Such grinding requires the removal of only a thin layer of metal—not more than the depth of corrugation and surface defects.

3.7.3 Ground Low Rail Profile

The ground low rail profiles should be designed to avoid contact positions that significantly face toward the field side. This is especially important for the low rails in curves, where hollow-worn wheels can give very high contact stresses on the field side. Fieldside contact can also increase the risk of rail rollover or gage spreading.

Varying low rail shape by grinding can alter the RRD by intentionally moving the contact positions on wheels to a desirable area, such as from the gage shoulder to the crown area by lowering the gage shoulder. Because the contact radius in the rail crown area is larger than on the rail shoulder, this adjustment reduces contact stress.

In designing ground rail profiles for curves, the contact conditions of both leading and trailing wheels should be considered. The leading wheels on the high rail are generally flanging for curves above 3 degrees. Vehicles that are designed with soft yaw suspension that allow the axles to steer may flange at higher degree curves. Therefore, it is recommended that the assessment of vehicle curving performance be conducted for the ground rail design. The trailing wheels are generally not flanging. They usually have a small lateral shift relative to the track center, depending on vehicle and track conditions.

3.7.4 Grinding Tolerance

During rail grinding, the transverse rail profile is produced by a series of straight facets from the individual grinding units. Thus, the grinding process unavoidably produces a polygon-curve approximation to the desired profile, and this causes variance between the actual ground rail profile and the target design rail profile. The stone pattern selections and settings can also cause deviations of ground rail shape from the target shape.

Thus, a grinding tolerance needs to be specified to limit the variation from the target shape. The tolerance is defined as the radial distance between the measurements of the ground rail profile and the target design profile. To check whether grinding has produced the design ground rail profile within the specified tolerance, the ground rail profile should be overlaid on the target rail profile.

Tolerances should be assessed as shown in Figure 3.24. Tolerance is evaluated from the highest point on the rail top

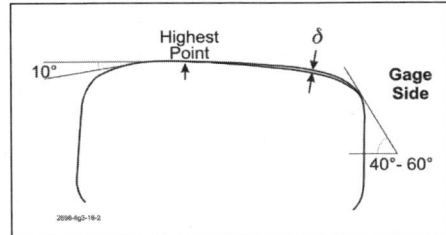

Figure 3.24. Example of grinding tolerance for a high rail.

to a point with an angle of 10 degrees on the field side and to a point with an angle of 40 to 60 degrees on the gage side. The angle on the gage side is based on the capacity of the grinders (the maximum angle that can be reached by the grinder).

Much work has been done in recent years to set tolerances on profiles given by rail grinding. Based on a survey of these references (14, 15, 16), the recommended tolerance for the ground rail transverse profile should be from −0.4 mm to +0.3 mm. Negative tolerances mean that the ground rail shape is below the design rail shape. Positive tolerances mean that the ground rail shape is above the design rail shape. The example in Figure 3.24 shows negative tolerances, that is, the measured rail profile is inside the template. Some grinders can reach even better accuracy with careful control of the grinding stone patterns.

Positive tolerances in the gage corner can be much more detrimental than negative tolerances. A large positive tolerance in the gage corner can lead to high contact stress and consequent high wheel- and rail-wear rates and the potential for crack formation. With good grinding accuracy, the ground rail shape will quickly wear to a profile that is conformal with the wheels passing over it.

Rail template gages are also commonly used to inspect the rail shapes during routine checks or after rail grinding. Experienced inspectors can estimate differences by looking at the gaps between the template and the actual rail shape (Figure 3.25). Note that in order to allow the gage to slide over the head of the two rails (even under the wide gage condition), the template gages usually do not have the whole shape of the rails.

3.7.5 Rail Lubrication after Grinding

Slight lubrication immediately after rail grinding can reduce the wheel flange climbing potential, just as lubrication after wheel truing can. The rough rail surface after grinding can reduce the limiting L/V ratio for flange climb.

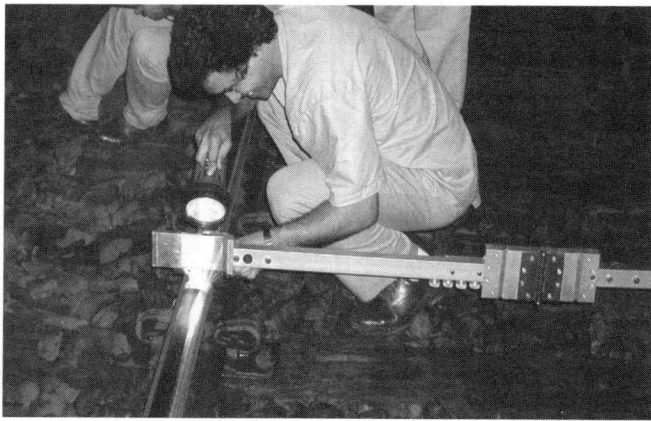 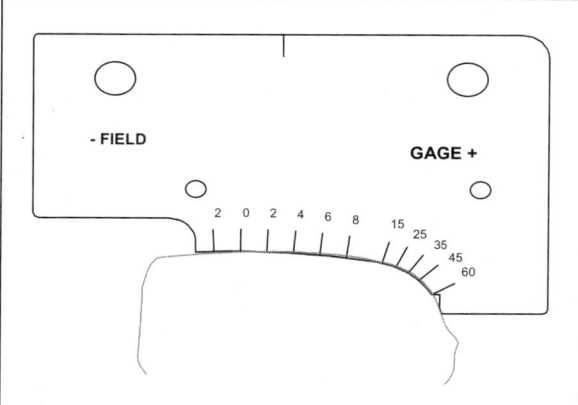

Figure 3.25. Example of ground rail template, "bar gage" and measurement.

3.8 EFFECT OF GAGE AND FLANGE CLEARANCE ON WHEEL/RAIL CONTACT

3.8.1 Effect of Rail Gage and Wheel Flange Clearance

Wheel flange clearance is the wheel lateral shift limit relative to the rail prior to wheel climb (Figure 3.26). It is directly related to rail gage, flange thickness, and wheel back-to-back spacing. Equation 3.3 computes the flange clearance (C_0) under the condition of designed gage and new wheel.

$$C_0 = \frac{G_s - B_s - 2f_s}{2} \quad (3.3)$$

where G_s and B_s are standard gage and wheel back-to-back spacing, respectively, and f_s is new wheel flange thickness.

Equation 3.4 computes the actual flange clearance (C) under the worn wheel/rail shapes and varied gage conditions.

$$C = C_0 + \frac{\Delta G}{2} + \Delta f = \frac{G_a - B_s - 2f_a}{2} \quad (3.4)$$

where

ΔG = the variation of rail gage from the standard value and can be both positive and negative based on the variation direction,

Δf = the variation of wheel flange thickness from the value for the new wheel, which is generally negative due to wear, and

G_a and f_a = the actual gage and flange thicknesses, respectively.

In general, the wheel back-to-back spacing is constant. It is obvious that the clearance increases with wide gage and a thin flange.

As discussed in Section 3.3.2.4, the flange clearance has an influence on the RRD in curves. Too narrow a gage can limit the RRD (and therefore the yaw displacement of the wheelset in curves), inducing wear, especially to high rails in curves. However, too wide a gage can increase the risk of gage widening derailment.

Figure 3.27 illustrates that the gage widening criterion is related to the wheel, rail geometries, and their relative positions.

When a wheel drops between the rails, as in Figure 3.28 the geometry of wheel and rail must meet the following expression:

$$G_a > B + W + f_a \quad (3.5)$$

where W is wheel width and B is the wheel back-to-back spacing.

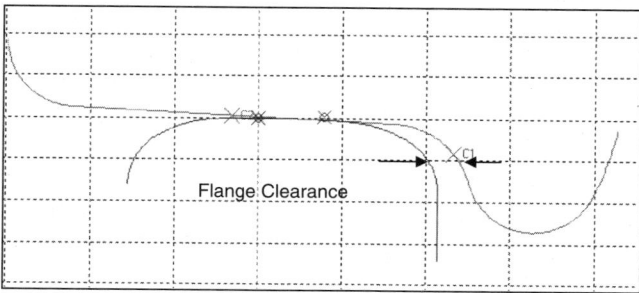

Figure 3.26. Illustration of flange clearance.

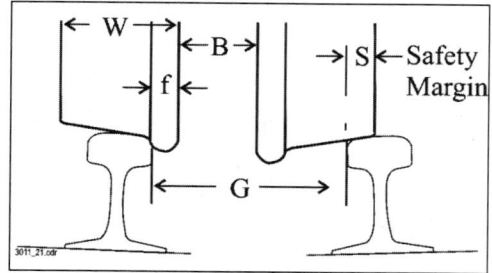

Figure 3.27. Wheel and rail geometry related to gage widening derailment.

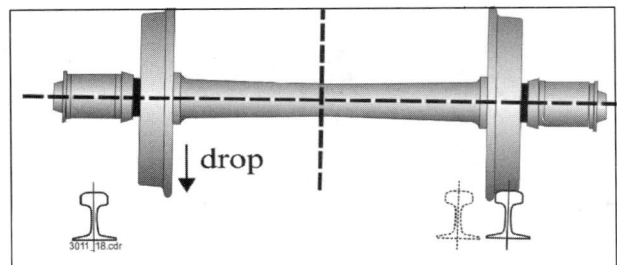

Figure 3.28. Derailment due to gage widening.

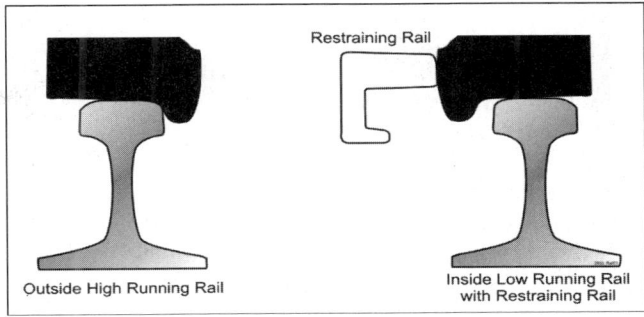

Figure 3.30. Restraining rail.

Therefore, a safety margin (*S*), expressed in Equation 3.6, represents the minimum overlap of wheel and rail required on the nonflanging wheel when the flanging wheel contacts the gage face of the rail. In this circumstance, the instantaneous flangeway clearance on the flanging wheel is zero.

$$(B + W + f_a) - G > S \quad (3.6)$$

where *G* is the gage spacing.

In general, the wheel back-to-back spacing (*B*) is a constant for a solid axle wheelset, and so is the wheel width *(W)*. However, different designs of wheel have different values for these two parameters. The flange thickness (f_a) is gradually reduced as the wheel wears. The track gage variations are influenced by multiple factors. Rail roll and lateral movement due to wheel/rail forces and weakened fasteners can widen the gage. Rail gage wear can also contribute to gage widening (Figure 3.29), which results in a gage wear of about 6 mm.

Widening the gage on sharp curves is a practice that has been adopted in some transit operations for improving vehicle curving. The limit to which the gage can be widened should be assessed carefully by considering the worst possible condition on that section of track, including both wheel and rail wear and wheel/rail lateral force. For example, a widened gage combined with hollow-worn wheels could cause rail roll or high contact stress.

3.8.2 Effect of Clearance between Low Rail and Restraining Rail

Restraining rails and guard rails have been frequently applied on sharp curves in transit operations to prevent flange climb derailment and to reduce high rail gage face wear. As illustrated in Figure 3.30, the restraining/guard rails are generally installed inside the low rail. In extremely sharp curves, restraining rails are sometimes installed on both inside and outside rails in order to also reduce low rail flange contact with the trailing wheelset.

The current practices of restraining rail installation vary, as shown in Table 3.4. The extension length of restraining rails at the ends of curves and the clearance to be set for the restraining rails may also vary within different transit systems.

The clearance between the low rail and the restraining rail is critical for the effectiveness of restraining rails. A clearance that is too tight reduces wheelset RRD required for truck curving while limiting flange contact on the high rail. Clearance that is too wide will cause a complete loss in the function of the restraining rail.

Wear at the wheel flange back and the contact face of the restraining rail can change the clearance between the low rail and the restraining rail. The wheel flange and high rail

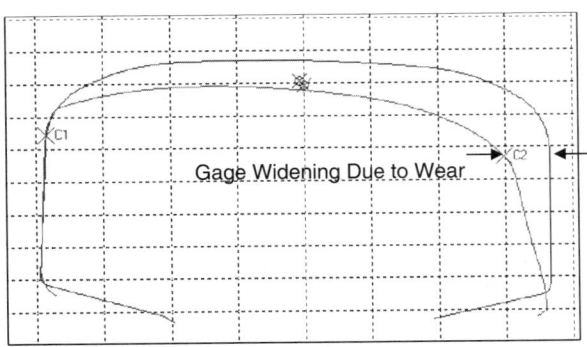

Figure 3.29. Gage widening due to wear.

TABLE 3.4 Examples of restraining rail installation

Transit System	Practice
Massachusetts Bay Transportation Authority (MBTA)	Restraining rail is installed on curves with a radius less than 1,000 ft.
Newark City Subway (Light Rail Line)	Restraining rail is installed on curves with a radius less than 600 ft.
Southeastern Pennsylvania Transportation Authority (SEPTA) (Rapid Transit Line)	Restraining rail is installed on curves with a radius less than 750 ft.
WMATA (Washington Metropolitan Area Transit Authority)	Restraining rail is installed on switches corresponding to less than 500-ft radius and curves with a radius less than 800 ft.
CTA (Chicago Transit Authority)	Restraining rail is installed on curves with a radius less than 500 ft.

gage wear can affect the amount of wheelset lateral shift in curves. Therefore, an installation clearance and a wear limit should be specified in order to maintain the vehicle curving performance in a desired range. Note that track lateral geometry irregularities including alignment and gage variations can also affect the performance of restraining rail.

No specific suggestions are given in this report regarding this clearance. The research team has proposed a study on this issue to further investigate the relation of wheel/rail force, clearance, and track curvature.

3.9 WHEEL/RAIL PROFILE MONITORING PROGRAM

A well-structured wheel/rail profile monitoring program can be an effective tool for detecting system performance and prioritizing maintenance needs.

3.9.1 Objectives of Profile Monitoring

The emphases of monitoring may differ for various systems depending on the existing vehicle and track conditions. For example, lower speed operations may pay more attention to vehicle curving behavior and wear issue since high speed instability is of no concern. Meanwhile, vehicle stability may become an issue for a higher speed operation.

The monitoring objectives can be defined into short-term and long-term objectives. The short-term objectives usually identify the problems that might be related to wheel/rail profiles or that the shapes of wheels and rails might provide some indication of, such as flange climb derailment potential, poor steering, and excessive wear.

The long-term objectives relate to system optimization, management, and maintenance. According to the performance trends and the wear patterns, a system level of improvement may be approached. For example, if an excessively high wear rate was observed on curves during rail profile monitoring, several related factors may be looked at, such as lubrication, wheel/rail profiles, restraining rail clearance, and the condition of vehicle suspension components. If wheel flange wear becomes excessive, as indicated by a thin flange, then track gage, vehicle curving performance, lubrication, and possibly the symmetry of wheel diameters on the same axle may need to be inspected. The sections that showed less satisfactory performances would get special attention.

Clearly setting the objectives of wheel/rail profile monitoring and prioritizing the emphases would make the monitoring program more effective and efficient.

3.9.2 Establish a Monitoring Program to Meet the Objectives

A wheel/rail profile monitoring program should define the following basic requirements:

- Measurement interval
- Measurement sample rate
- Distribution of measurement sites
- Measurement accuracy requirement
- Required measurement devices
- Documentation procedures
- Analysis procedures
- Reporting procedures

Depending on the objectives, more detailed descriptions can be included in the program.

Note that the measurement interval should be set based on the loading and operation frequency, so as to correctly reveal the trend, but not so short that it would increase the monitoring cost. The sample rate should be sufficient to provide information that would be representative of the wheel/rail populations.

For monitoring rails, key locations in the system should be marked. Here the measurements should be performed at exactly the same locations in order to accurately determine the changes of profile due to wear. Wheels selected for monitoring should be marked to trace the profile changes.

3.9.3 Integrate the Profile Monitoring Program into Vehicle/Track Maintenance Program

As discussed previously, results of wheel/rail interaction are not only related to wheel/rail profiles. They are also affected by the vehicle and track conditions in that system, most often vehicle suspensions, track geometries, and lubrication. In many cases, the improvement of vehicle performance or wheel/rail wear relies on the combined improvement in wheel/rail profiles, track maintenance, and lubrication. Therefore, the profile monitoring program should be integrated into the vehicle/track monitoring/maintenance program.

CHAPTER 4

GLOSSARY OF TECHNICAL TERMS

The following are the technical terms commonly used in the analysis of wheel and rail interaction. They are listed here to help the reader clarify the physical meaning of each term. Some of them may have been explained in the different sections of the report. They are again listed here for the convenience of reference.

Contact Angle	The angle of the plane of contact between the wheel and rail relative to the track plane. This should not be confused with the flange angle described below.
Contact Stress	The force acting per unit area of the contact surface. See Section 3.3.
Conicity	The ratio of RRD between the left and right wheel over the wheelset lateral displacement: $$\lambda = \frac{r_L - r_R}{2y}$$ where r_L and r_R are the rolling radius of the left and right wheel, respectively, and y is the wheelset lateral displacement.
Flange Angle	The angle between the tangential line of the point on the wheel flange and a horizontal line. The maximum wheel flange angle (δ) is defined as the angle of the plane of contact on the flange relative to the horizontal (see Figure 3.1), This should not be confused with the contact angle described above.
Flange Back Clearance	The distance between the flange back face of the wheel and the contact face of the restraining rail.
Flange Clearance	Wheel flange clearance is the wheel lateral shift limit relative to rail prior to wheel climb. See Figure 3.26.
Flange Length	The total length of the arc or line on the flange of the wheel profile, starting from the point where the maximum flange angle begins to the point where the angle on the flange tip reduces to 26.6 degrees. See Figure 3.2.
Gage	The distance between the left and right rail measured from the gage point on the rail. The gage point is defined at a specific height below the rail top. Different distances are used by various transit properties. See Figure 3.27.
Independently Rotating Wheels	A pair of wheels on a common axle that rotate independently of each other without any relative rotational constraint. See Chapter 2 of Appendix B.
L/V Ratio	The ratio of lateral-to-vertical (L/V) wheel/rail contact forces. Sometimes referred to as the Y/Q ratio.
Rolling Contact Fatigue	The deformation and damage on a wheel or rail caused by the repetitive experience of excessive normal and tangential forces.
Rolling Radius Difference (RRD)	The difference in rolling radius between the contact points on the two wheels of a common wheelset. See Figure 3.14.
Wheel Back-to-Back Distance	The distance between the flange back plane of the left and right wheel. See Figure 3.27.

Wheelset Angle-of-Attack (AOA)   The angle (ψ) between the axis of rotation of the wheelset and a radial line in a curve or a line perpendicular to the track centerline on tangent track, as shown in Figure 4.1.

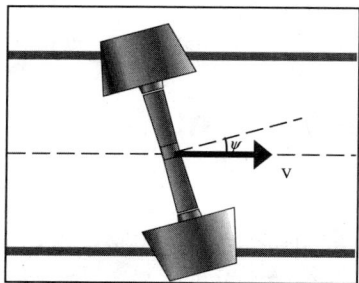

Figure 4.1.   Wheelset AOA.

Wheelset Lateral Shift   The lateral displacement of the wheelset center relative to the track centerline.

REFERENCES

1. Nadal, M. J., *Locomotives a Vapeur*, Collection Encyclopedie Scientifique, Biblioteque de Mecanique Appliquee et Genie, Vol. 186, Paris, France, no date.
2. Wu, H., and J. Elkins, "Investigation of Wheel Flange Climb Derailment Criteria," Report R-931, Association of American Railroads, Washington, D.C., July 1999.
3. *Manual of Standards and Recommended Practices*, Section G, "Wheels and Axles," Association of American Railroads, Washington, D.C., 1998.
4. *Field Manual of AAR Interchange Rules*, Association of American Railroads, Washington, D.C., 2004.
5. Sawley, K., C. Urban, and R. Walker, "The Effect of Hollow-Worn Wheels on Vehicle Stability in Straight Track," Proceedings, Contact Mechanics and Wear of Rail/Wheel System, the 6th International Conference, Gothenburg, Sweden, June 2003.
6. Sawley, K., and H. Wu, "The Formation of Hollow-Worn Wheels and Their Effect on Wheel/Rail Interaction," Proceedings, Contact Mechanics and Wear of Rail/Wheel System, the 6th International Conference, Gothenburg, Sweden, June 2003.
7. Wu, H., and K. Sawley, "New Solutions to Familiar Problems," *International Railway Journal*, pp. 40-41, December 2003.
8. Wilson, N., et al., *NUCARS Users Manual, Version 2.3*, Transportation Technology Center, Pueblo, Colo., 2003.
9. International Heavy Haul Association, *Guidelines to Best Practices for Heavy Haul Railway Operations: Wheel and Rail Interface Issues*, Virginia Beach, Va., May 2001.
10. Evans, J., and M. Dembosky, "Investigation of Vehicle Dynamic influence on Rolling Contact Fatigue on UK Railways," *The Dynamics of Vehicles on Roads and on Tracks*, Volume 41, 2003.
11. Parsons Brinckerhoff Quade & Douglas, Inc., *TCRP Report 57: Track Design Handbook for Light Rail Transit*, TRB, National Research Council, Washington, D.C., 2000.
12. *APTA Passenger Rail Safety Standard Task Force Technical Bulletin*, 1998-1, Part 1, American Public Transportation Association, 1998.
13. *Manual for Railway Engineering*, Volume 1, Chapter 4, American Railway Engineering and Maintenance of Way Association, Landover, Md., 2003.
14. Ishida, M., and N. Abe, "Experimental Study on the Effect of Preventive Grinding for Shinkansen Rails," 6th International Heavy Haul Conference, Cape Town, South Africa, April 1997.
15. Grassie, S., "Requirements for Transverse Railhead Profile and Through Rail Grinding," 6th International Heavy Haul Conference, Cape Town, South Africa, April 1997.
16. Worth, A., J. Hornaday, and P. Richards, "Prolonging Rail Life Through Rail Grinding," 3rd International Heavy Haul Conference, Vancouver, British Columbia, 1986.

APPENDIX A:

Effect of Wheel/Rail Profiles and Wheel/Rail Interaction on System Performance and Maintenance in Transit Operations

EFFECT OF WHEEL/RAIL PROFILES AND WHEEL/RAIL INTERACTION ON SYSTEM PERFORMANCE AND MAINTENANCE IN TRANSIT OPERATIONS

SUMMARY

The research team performed a survey of representative transit systems to identify the common problems and concerns related to wheel/rail profiles in transit operations. This survey was conducted as part of Phase I of this project to develop wheel/rail profile optimization technology and flange climb criteria.

The research team conducted onsite surveys at six representative transit systems that involve both light rail and rapid transit operations to collect information related to wheel/rail profiles and wheel/rail interactions. Several vehicle maintenance shops and track sites were visited to observe current wheel/rail profile related practices and problems. Summaries of the information from five of the systems visited are included in the Appendixes A-1 through A-5.

The survey identified the following common problems and concerns related to wheel/rail profiles and wheel/rail interaction in transit operation:

- Adoption of low wheel flange angles can increase the risk of flange climb derailment. High flange angles above 72 degrees are strongly recommended to improve operation safety.
- Rough wheel surface finishes from wheel re-profiling can increase the risk of flange climb derailment. Final wheel surface finish improvement and lubrication could mitigate the problem considerably.
- Introduction of new wheel and rail profiles need to be carefully programmed for both wheel re-profiling and rail grinding to achieve a smooth transition.
- Without adequate control mechanisms, independently rotating wheels can produce higher lateral forces and higher wheel/rail wear on curves.
- Cylindrical wheels may reduce the risk of vehicle hunting, but can produce poor steering performance on curves.
- Some wheel and rail profile combinations used in transit operations have not been systematically evaluated to ensure they have good performance on both tangent track and curves under given vehicle and track conditions.
- Severe two-point contact has been observed on the designed wheel/rail profile combinations from several transit operations. This type of contact tends to produce

poor steering on curves, resulting in higher lateral force and a higher rate of wheel/rail wear.
- Track gage and restraining rails need to be carefully set on curves to allow sufficient RRD and to reduce some high rail wear and lateral force.
- Wheel slide and wheel flats occur on several transit systems, especially during the fall season. Although several technologies have been applied to mitigate the problem, transit operators are in need of more effective methods.
- Generally, noise related to wheels and rails is caused by wheel screech/squeal, wheel impact, and rail corrugations. Wheel/rail lubrication and optimizing wheel/rail contact could help to mitigate the noise problems.
- Wheel/rail friction management is a field that needs to be further explored. Application of wheel/rail lubrication is very limited in transit operation due to the complications related to wheel slide and wheel flats.
- Reduction of wheel/rail wear can be achieved by optimization of wheel/rail profiles, properly designed truck primary suspension, improvement of track maintenance, and application of lubrication.
- Without a wheel/rail profile measurement and documentation program, transit operators will have difficulty reaching a high level of effectiveness and efficiency in wheel/rail operation and maintenance.
- Further improvement of transit system personnel understanding of wheel/rail profiles and interaction should be one of the strategic steps in system improvement. With better understanding of the basic concepts, vehicle/track operation and maintenance activities would be performed more effectively.

CHAPTER 1
INTRODUCTION

This project included two phases (Table A-1).

This report describes the methodology and engineering behind wheel/rail profile optimization and the results derived from the work performed in Task 1, Phase I of the program.

1.1 BACKGROUND

A railroad train running along a track is one of the most complex dynamic systems in engineering due to the many nonlinear components. The interaction between wheels and rails is an especially complicated nonlinear element of the railway system. Wheel and rail geometry—involving cross section profiles, geometry along the direction of travel, and varying shapes due to wear—has a significant effect on vehicle dynamic performance and operating safety.

Transit systems are usually operated in dense, urban areas, which frequently results in lines that contain a large percentage of curves, or curves with small radii, which can increase wheel and rail wear and increase the potential for flange climb derailments. Transit systems also operate a wide range of vehicle types, such as those used in commuter rail service, heavy or rapid transit and light rail vehicles, with a wide range of suspension designs and vehicle performance characteristics.

In general, transit systems (in particular, light rail and subway systems) are locally operated. Without the requirement of interoperability, many transit systems have adopted different wheel and rail profile standards for different reasons. Some of these standards are unique to a particular system. Older systems frequently have wheel and rail profile standards that were established many years ago. For some older systems, the reasons that specific profiles were adopted have been lost in time. Newer systems have generally selected the wheel/rail profiles based on the increased understanding of wheel/rail interaction.

Increasing operating speed and introducing new designs of vehicles have further challenged transit systems to maintain and improve wheel/rail interaction. Good overall performance can be achieved by optimizing vehicle design, including suspension and articulation, to work with optimized wheel and rail profiles. However, possibilities of modifying existing vehicles are limited. Along with other activities, optimization of wheel/rail contact is one of the strategies for maintaining or improving vehicle performance.

1.2 OBJECTIVES

There are two main objectives of wheel/rail profile assessment technology development:

- Identify common problems and concerns related to wheel/rail profile and interaction in transit operations.
- Provide guidelines to transit system operators for wheel/rail profile assessment, monitoring, and maintenance.

This appendix fulfills the requirement of the first objective.

TABLE A-1 Tasks in the program of wheel/rail profile optimization technology and flange climb criteria

Program Tasks		
Phase I	Task 1	Define common problems and concerns related to wheel/rail profiles in transit operation
	Task 2	Propose preliminary flange climb criteria for application to transit operation
Phase II	Task 1	Develop general guidelinesfor wheel/rail profile assessment applied to transit operation
	Task 2	Propose final flange climb derailment criteria validated by the test data

CHAPTER 2

METHODOLOGY

To develop wheel/rail profile assessment technology, the existing problems and concerns related to wheel and rail profiles in transit operations first need to be identified. A survey has been conducted of selected transit systems to examine the current state of common practices in wheel/rail operation and maintenance.

2.1 SELECTION OF SYSTEMS FOR SITE VISITS

The research team compiled a partial list of transit systems based on the 2003 *Membership Directory of American Public Transportation Association (1)* and the research team's knowledge of these systems, as shown in Table A-2. These systems operate a large number of cars and have a variety of types of operation. They are mainly located in four geographic areas:

- Washington, D.C.—Baltimore
- Chicago
- California
- the Northeast (Boston–New York–Philadelphia)

The research team visited several of these transit agencies to perform the survey. Due to budget limitations, the team sur-

TABLE A-2 List of large transit systems

	Transit System	Light Rail Cars	Bi-level	Rapid Transit	Commuter Coach	Locomotive	Total	Geographic area
1	San Francisco Bay Area Rapid Transit (BART)			669			669	California
2	San Francisco Municipal Railway	136					136	California
3	Southern California Regional Rail Authority (Los Angeles)				146	37	183	
4	Metropolitan Atlanta Rapid Transit Authority			238			238	Atlanta
5	Chicago Transit Authority			1190			1190	Chicago
6	Chicago Metra		781		165	139	1085	
7	Massachusetts Bay Transportation Authority	185		408	362	80	1035	Northeast
8	New Jersey Transit Corporation (NJ TRANSIT)	45			844	139	1028	
9	Port Authority Transit Corporation (Lindenwold, NJ)	121					121	
10	Metropolitan Transportation Authority (New York)			8231			8231	
11	Port Authority Trans-Hudson Corporation (New York)				342		342	
12	Southeastern Pennsylvania Transportation Authority	197		345	349		891	
13	Maryland Transit Administration	53		100	110	30	293	Washington D.C./Baltimore
14	Washington Metropolitan Area Transit Authority			882			882	

TABLE A-3 Transit systems visited during the survey

	Area	Agencies Visited
Trip 1	Northeast	• MBTA • New Jersey Transit Corporation (NJ TRANSIT)
Trip 2	Washington, D.C. and Philadelphia	• Washington Metropolitan Area Transit Authority (WMATA) • SEPTA
Trip 3	Chicago area	• Chicago Transit Authority (CTA) • Chicago Metra (not included in the summaries of site visits)

veyed six systems in three main geographic areas (Table A-3). These six systems are considered representative of transit operations in North America.

2.2 SITE VISIT

During each site visit, the following information was researched:

- Current problems related to the wheel/rail profiles in that system.
- Historical information of wheel/rail related problems.
- Map of route, curve distribution, and operating speed.
- New wheel/rail profile designs.
- Worn wheel/rail profiles, if available.
- Wheel/rail wear historical data.
- Vehicle design information.
- Rail lubrication practices.
- Current wheel/rail maintenance practice.
- Any other wheel/rail related problems.

Results from the individual surveys are contained in the Appendixes A-1 to A-5.

2.3 ANALYSIS OF SURVEY INFORMATION

The research team has carefully studied the information from the survey. Analysis has been performed with an emphasis on the wheel/rail profiles and issues related to wheel/rail interaction. These issues include the following:

- Typical wheel/rail profiles used in transit operations.
- Wheel/rail contact patterns.
- Wheel/rail wear patterns.
- Safety concerns related to wheel/rail profiles.
- Vehicle curving performance and lateral stability behavior as affected by wheel/rail profiles.
- Other issues related to wheel/rail profiles.

2.4 IDENTIFYING COMMON WHEEL/RAIL PROFILE ISSUES

Based on the survey and the survey information analysis, the common problems and concerns related to wheel/rail profiles in transit systems were identified and are further summarized in this appendix. Based on the survey, there is a clear understanding of what guidelines transit operators need related to wheel/rail profiles.

CHAPTER 3
TRANSIT SYSTEM SURVEY

As a way of focusing expectations, prior to each visit to the representative transit systems, a questionnaire was sent to a primary contact person at each location focusing on the following topics:

- Existence and type of wheel/rail profile related problems on the system.
- Remedies tried (successfully or not) for the wheel/rail problems.
- Determining whether the problems are specific to particular vehicle types and/or track locations.
- Opinions of existing track and/or car conditions versus design and maintenance standards.
- Lubrication practices.
- Employee wheel/rail interface training needs.
- Existence of a wheel re-profiling program.
- Existence of, and criteria for, a rail grinding program.
- Percentage of budget spent on various facets of wheel/rail maintenance.
- Any other vehicle performance research.
- Driving factors behind the wheel/rail work (e.g., economics, safety).
- Major corrections needed to improve wheel and rail interaction.

Each visit began with a presentation on the effect of wheel/rail interaction on vehicle performance to groups that included track maintenance, vehicle maintenance, and operating personnel. This presentation (approximately 1 hour long) was intended to improve the group's understanding of wheel/rail contact systems and the importance of wheel/rail profile optimization and to stimulate discussion of wheel/rail issues. Five major topics were included in the presentation:

- Fundamentals of wheel and rail contact.
- Vehicle dynamics related to wheel and rail shapes.
- Problems caused by incompatible wheel/rail profiles.
- Wheel/rail lubrication.
- Wheel/rail maintenance.

After the initial seminar, a group discussion was held to explore wheel/rail topics. With the involvement of personnel from operations, track maintenance, and vehicle maintenance, many problems and concerns were reviewed at a systems level. The discussion continued with individual interviews, during which the research team collected much information related to practice, standards, and rules. At each site visited, a wheel/axle/car shop tour was conducted with a primary mechanical representative. Then, an on-track visit and discussion was performed with an engineering representative.

Appendixes A-1 to A-5 provide a brief summary of the information from each individual system visited. These summaries have been reviewed by the relevant systems to ensure the accuracy of the information. The topics in each summary include the following:

- Wheel and rail profiles.
- Wheel life and wheel re-profiling.
- Rail life and rail grinding.
- Track standards.
- Fixation methods.
- Lubrication and wheel slide.
- Noise.
- Major concerns and actions.

CHAPTER 4

COMMON PROBLEMS OR CONCERNS RELATED TO WHEEL/RAIL PROFILES

The common problems and concerns discussed in this section are summarized from the survey information. They may not apply to every transit system. An issue is addressed here because it was of interest or was considered by more than one system and because it falls into the general area of wheel/rail interaction.

In this section, the following issues related to wheel/rail profiles and wheel/rail interactions are discussed:

- Wheel flange angle.
- Surface finish from wheel re-profiling.
- System transition in increasing wheel flange angle.
- Independently rotating wheels.
- Cylindrical tread wheels.
- Wheel/rail contact condition analysis.
- Track gage and flangeway clearance.
- Wheel slide and wheel flats.
- Noise.
- Rail lubrication.
- Wheel/rail wear.
- Wheel/rail profile monitoring and documentation.

A brief description of the theories related to each issue is provided for a better understanding on the cause of the problem and the damage that might result.

4.1 WHEEL FLANGE ANGLE

The maximum flange angle of the designed wheel profiles applied in transit operation ranges between 63 and 75 degrees. Table A-4 lists the wheel flange angles received for the six visited systems.

It was noticed that the wheel profile drawings received from some systems have no direct measure of wheel flange angle. Some flange angles listed in Table A-4 were obtained by converting the wheel profile drawings received to CAD drawings. Then the flange angles were accurately derived from the CAD drawings. During the survey, when asked about the flange angle, the engineers in the vehicle maintenance group usually would only reference the drawings but not know the actual angle if there was no direct measure of flange angle in the drawing.

The maximum wheel flange angle (δ) is defined as the angle of the plane of contact on the flange relative to the horizontal (Figure A-1), and it has a significant effect on wheel flange climb derailment. Figure A-1 illustrates the system of forces acting on the flange contact point. Lateral force (L) and vertical force (V) are exerted on the rail by the wheel. Reacting forces exerted on the wheel by the rail are the normal force (F_3) and the lateral creep force (F_2) in the plane of contact.

TABLE A-4 Maximum wheel flange angle of designed wheels (the blank indicates no such service in that system)

System	Light Rail Cars Flange Angle	Rapid Transit Cars Flange Angle	Commuter Cars Flange Angle
MBTA	63 degrees (in the transition to 72 degrees)	No information was received	No information was received
NJ TRANSIT	75 degrees		72 degrees
SEPTA	60-65 degrees (in specified tolerance)	63 degrees	72 degrees
WMATA		63 degrees	
Chicago Metra			75 degrees
CTA		68 degrees	

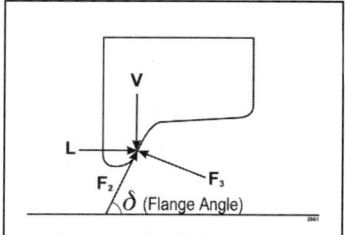

Figure A-1. Flange forces at wheel climb.

Equating forces in the lateral and vertical directions give the following equation: *(2)*

$$\frac{L}{V} = \frac{\tan \delta - F_2/F_3}{1 + F_2 \tan \delta / F_3} \quad (A\text{-}1)$$

This equation gives the minimum L/V ratio at which flange climb derailment can occur for any value of F_2/F_3 at a specified maximum contact angle. Nadal's criterion *(3)*, proposed in 1908 and still used extensively for derailment assessment, can be derived from Equation A-1 for the saturated condition of $\frac{F_2}{F_3} = \mu$ where μ = the coefficient of friction between the wheel and the rail (see Equation A-2):

$$\frac{L}{V} = \frac{\tan \delta - \mu}{1 + \mu \tan \delta} \quad (A\text{-}2)$$

If the maximum contact angle is used, Equation A-2 gives the minimum wheel L/V ratio at which flange climb derailment may occur for the given contact angle and friction coefficient μ. In other words, below this L/V value, flange climb cannot occur. Figure A-2 plots the relation of limiting L/V ratio and maximizing flange angle at different levels of friction coefficient between wheel and rail.

Figure A-2 gives two examples of wheel flange angles. One is the AAR-1B wheel profile with a 75-degree flange angle, and the other wheel profile has a 63-degree flange angle. At a friction coefficient of 0.5, which is the dry wheel/rail contact condition, the limiting L/V value is 1.13 for wheel profiles with a 75-degree flange angle (such as the AAR-1B wheel) and 0.73 for the wheels with a 63-degree flange angle. Clearly, wheels with low flange angles have a higher risk of flange climb derailment.

Increasing the design wheel flange angle to reduce the risk of flange climb derailment has been a common practice for transit systems. Due to historic reasons, some older transit systems have adopted relatively low wheel flange angles in the range of 63 to 65 degrees. The low flange angles are prone to flange climb derailment and have less compatibility with different truck designs. Newer transit systems generally start with a wheel profile having a flange angle of 72 to 75 degrees.

A wheel profile with a higher flange angle can reduce the risk of flange climb derailment and can have much better compatibility with any new designs of vehicle/truck that may be introduced in the future compared to wheels with lower flange angles. Also, with higher L/V ratio limits (according to the Nadal flange climb criterion), high flange angles will tolerate greater levels of unexpected track irregularity.

4.1.1 Derailments of Low Floor Light Rail Vehicles Due to Low Flange Angle

Figure A-3 compares two examples of designed wheel profiles used by transit systems. First is a wheel profile with a flange angle of 63 degrees that was previously applied to all vehicles on MBTA's Green Line (light rail), including new Number 8 cars. The second example is the profile applied to the NJ TRANSIT's Newark city subway (light rail) with a flange angle of 75 degrees.

As shown in Figures A-4 and A-5, MBTA's Number 8 cars have a structure similar to that of the NJ TRANSIT light rail cars (LRVs) with the low-platform level boarding and low floor for handicapped accessibility. These types of cars have three sections and double articulation at the center unit. The center unit is equipped with independent rotating wheels.

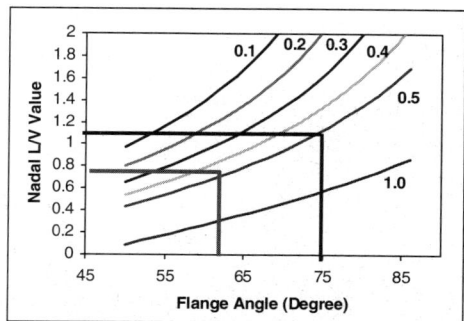

Figure A-2. Relationship of limiting wheel L/V ratio and maximum flange angle.

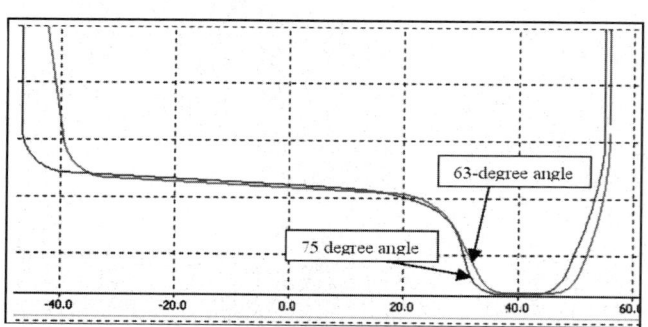

Figure A-3. Examples of designed wheel profiles.

Figure A-4. The Number 8 car of MBTA Green Line.

The cars from these two light rail systems show different dynamic performance due to the differences in suspension and wheel profile design. However, the 63-degree wheel flange angle, combined with other track and vehicle situations, apparently contributed to derailments of the Number 8 cars in 2000 and 2001. One of MBTA's remedial actions has been to increase the wheel flange angle from 63 degrees to 75 degrees by introducing a new wheel profile. Rail grinding has also been performed to reshape the rail gage corner to help the wheels maintain the 75-degree flange angle. Combined with other improvements in track maintenance, the derailment of the Number 8 cars due to the low flange angle has been eliminated.

In comparison, there are no derailment concerns with the similar cars on the NJ TRANSIT subway system. The wheels in the cars were designed with a 75-degree flange angle.

4.1.2 Derailment of Rapid Transit Vehicles Due to Low Flange Angle

To reduce wheel and rail wear, WMATA adopted the British worn tapered wheel profile in 1978 to replace the old cylindrical profile. This wheel profile has a 63-degree flange angle. In 1993, six low speed flange climb derailments occurred on curves in the yards. Guardrails were installed later in those derailment locations. In August 2003, a flange climb derailment occurred on a service train. Among other causes, the consultant for the derailment investigation has suggested that the 63-degree flange angle may have increased the risk of flange climb derailment. WMATA has been considering the improvement of wheel profile to a larger flange angle of 72 to 75 degrees.

4.1.3 Additional Transit System Wheel Profile Designs

Among the 14 wheel profile drawings of U.S./North American light rail systems that are included in the *Track Design Handbook for Light Rail Transit* (excluding SEPTA and MBTA wheel profiles, which have been discussed in Table A-4), 8 have no direct measures of flange angle, and 2 of the remaining 6 have a design flange angle of 63 degrees (*4*).

This handbook proposed a wheel flange angle of 70 degrees based on Heumann's design. The APTA's *Passenger Rail Safety Standard Task Force Technical Bulletin* (*5*) provides guidance on reducing the probability of wheel-climb derailment, suggesting a minimum wheel flange angle of 72 degrees (suggested tolerances are +3.0 degrees and −2.0 degrees).

4.2 WHEEL RE-PROFILING

4.2.1 Rough Surface from Wheel Re-Profiling

Wheel truing is a process that re-profiles the wheel shape and removes surface defects such as flats, spalls, and shellings. Two types of wheel re-profiling machines are commonly used. Figure A-6 shows the milling type, which has a cutting head with many small cutters. The arrangement of the cutters forms the wheel profile. Figure A-7 shows the lathe type, which has a wheel profile template; the single cutter cuts the wheel by following the shape of a template.

Figure A-5. NJ TRANSIT LRV.

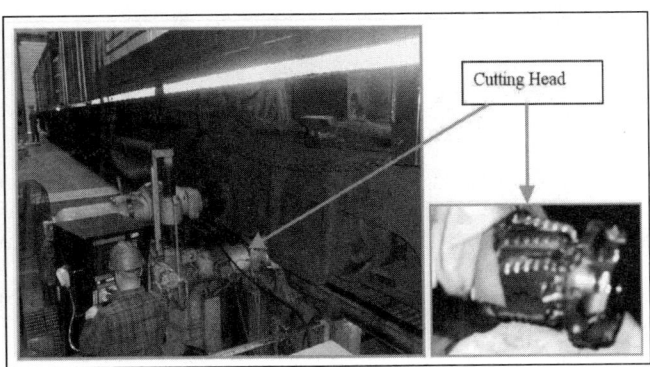

Figure A-6. Milling type wheel re-profiling machine.

Figure A-7. Lathe type wheel re-profiling machine.

Several systems have reported flange climb derailments occurring at curves or switches in yards when the cars were just out of the wheel re-profiling machines. This type of derailment was likely caused by the wheel surface roughness after wheel re-profiling. Figure A-8 compares the wheel surfaces just after re-profiling and the surface after many miles of running. The left wheel in Figure A-8 was re-profiled by the milling type machine with very clear cutting traces on the surface. The middle wheel was re-profiled by a lathe type machine with shallower cutting traces. The right wheel was returned to the shop from service with a smooth surface but had a flat spot on the tread.

Generally, the coefficient of friction for dry and smooth steel-to-steel contact is about 0.5. The effective coefficient of friction for rough surface condition can be much higher. For example, if the friction coefficient reaches 1.0, the L/V limit would be 0.5 for a 75-degree flange angle and 0.3 for a 63-degree flange angle (as shown in Figure A-2). Therefore, the rough surface produced by wheel re-profiling could significantly reduce the L/V limit for flange climb. The low flange angle further increases the derailment risk.

Several remedies may improve the surface condition:

- Frequently inspecting the cutting tools—especially for the milling type machine. Dulled tools can produce a very rough surface. Sometimes the grooves on the wheels were obvious.
- Addressing the final surface tuning. In this step, there is no significant material removal but rather a light cut for smoothing the surface. WMATA has included this step in its wheel re-profiling procedures.

Further, lubrication after re-profiling can be an effective way to prevent flange climb derailment on newly re-profiled wheels. Again, referring to Figure A-2, reducing the friction coefficient at wheel/rail interface can increase the L/V limit for flange climb. The sharp asperities on the wheel surface after re-profiling may quickly deform or wear off in operation due to very high locally concentrated contact stress. After some

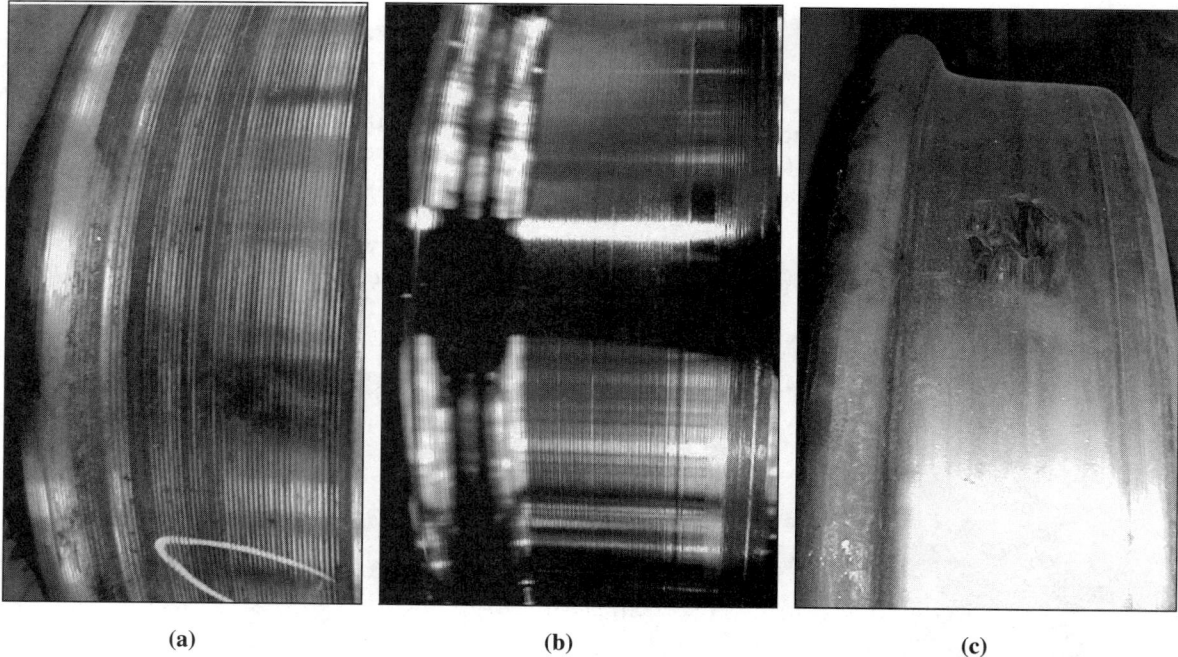

Figure A-8. Comparison of wheel surface roughness. (a) Surface after wheel re-profiling from milling type machine, (b) Surface after wheel re-profiling from lathe type machine, and (c) surface of wheel back from operation with a flat spot.

distance of operation, the wheel surface should be smoother. Light lubrication can help wheels safely pass this rough-to-smooth transition. WMATA now manually lubricates all wheels immediately after re-profiling. CTA has installed wayside lubricators on the curves as well as guardrails in their yards.

4.2.2 Wheel Diameter Difference after Re-Profiling

Improper setting of tools or wheelset position during re-profile can cause a diameter difference in wheels on the same axle. The difference in diameter between two axles in the same truck can be caused by varying material removal rates.

The difference in diameter for the wheels in the same (coupled) axle could lead either to the truck running offset if two axles in the same truck have a similar pattern of diameter difference or to unstable truck performance if only one of the axles has the diameter difference. In this situation, one axle tends to drag the truck to an offset position while the other axle tends to pull the truck back to the track center. The situation will be even more complicated if both axles have differences in diameter but with different patterns. Table A-5 lists the diameter tolerances used in several transit systems.

4.2.3 Introducing New Wheel/Rail Profiles

Introduction of a new wheel profile or a new rail profile will require a transition period to bring the wheel/rail system into equilibrium.

Wheels and rails in a system generally wear into a conformal, stable state in terms of the profile shapes. If either wheel or rail profile needs to be redesigned and the new profile has a significantly different shape compared to the existing profile, the existing wheel/rail conformality will be lost. A program needs to be carefully designed for a smooth transition to reach a new equilibrium of wheel/rail contact.

As discussed in Section 4.1, several transit systems have adopted wheel profiles with relatively low flange angles due to historic reasons. To reduce the risk of flange climb and increase safety margins, some systems have started or intend to adopt new wheel profiles with higher flange angles. If only wheel profiles are changed, the initial situation of a new wheel with higher flange angle contacting with an existing worn rail with lower gage angle would be likely (see Figure A-9). Because of the different angles at the wheel flange and the rail gage, the contact position at the flange is likely to be low and contact stress is likely to be high due to the small contact area. The 75-degree angle at the contact position shown in Figure A-9 will very likely wear into a lower angle (63 to 75 degrees). It may also result in severe two-point contact on curves and adversely affect truck steering.

MBTA Green Line (light rail) faced this problem. The wheel profile used on both existing Number 7 cars and new Number 8 cars originally had a flange angle of 63 degrees. To reduce the risk of flange climb derailment on the Number 8 cars, MBTA Green Line implemented an Interim Wheel Profile (IWP) with a flange angle of 75 degrees for the Number 8 cars. However, the transition from the old wheel profile to the new wheel profile with higher flange angle did not go smoothly on the Green Line. Due to the capacity of wheel re-profiling and rail grinding, many Number 7 cars (115 cars), the majority of the Green Line car fleet, were still equipped with wheels with a 63-degree flange angle (or slightly higher at worn condition) and some sections of worn gage face rail were still in use (with a gage angle of 63 degrees or slightly higher). These low existing angles on wheel flanges and rail gage faces continued to resist the profile transition and cause fast wear on the flanges of the few cars (10 to 20 cars) with 75-degree-flange-angle wheels. This high wear rate required very frequent re-profiling of the IWP wheels (as little as 3,000 operating miles between wheel re-profiling) in order to maintain the desired 75-degree flange angle.

TABLE A-5 Examples of currently used diameter tolerance after wheel re-profiling

System	Diameter Tolerance after Wheel Re-profiling
SEPTA	1/8 in. within the same axle
	1/4 in. axle-to-axle in the same truck
	1/2 in. truck-to-truck in the same car
WMATA	1/16 in. within the axle
	1/4 in. axle to axle in the same truck
	1/2 in. truck to truck in the same car
Chicago Metra	1/8-in. variation left-to-right within an axle
	1/4 in. axle to axle within a truck
	1/4 in. truck to truck within a car
CTA	3/64 in. within an axle (0.046 in.)
	1 in. axle to axle in the same truck
	1 in. truck to truck within the same car

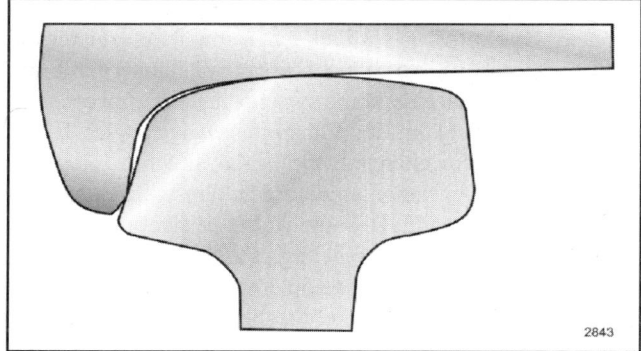

Figure A-9. Contact condition at wheel flange.

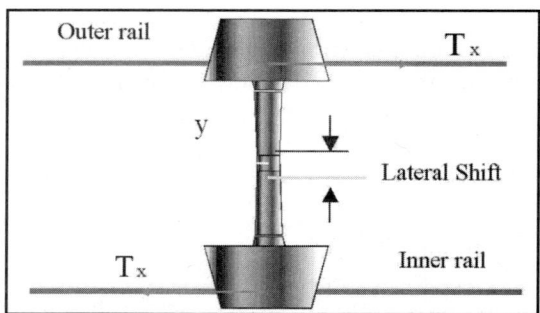

Figure A-10. Steering moment produced by longitudinal forces.

Therefore, when changing wheel (or rail) profile is necessary, a transition program should be carefully planned by considering the capacity of both wheel re-profiling and rail grinding. If the profile change is significant, one or more interim profiles may be needed to gradually approach the desired profile.

4.3 INDEPENDENTLY ROTATING WHEELS

Independently rotating wheels are generally used in Light Rail low-floor cars. Low-floor cars are used for the advantages of easier boarding and handicapped accessibility. Although only a few systems currently operate cars with independently rotating wheels, this issue is still worth discussion in consideration of introducing these types of vehicles in the future.

Wheels mounted on a solid axle must move at the same rotational speed. To accommodate running in curves, a taper is usually applied to the wheels. The wheelset shifts sideways, as shown in Figure A-10, to allow the outer wheel to run with a larger rolling radius than the inner wheel. The resulting longitudinal creep forces at the wheel/rail interface for wheels on the same axle form a moment that steers the truck around the curve. Previous flange climb studies indicate that as the ratio of longitudinal force and vertical force increases, the wheel L/V ratio required for derailment also increases (2). This is illustrated in Figure A-11. Therefore, the Nadal flange climb criterion can be relaxed based on the level of longitudinal force. The flange climb would occur at an L/V ratio above the Nadal limiting value when there is a significant longitudinal force.

Independently rotating wheels do not produce longitudinal forces on curves to form the steering moment (Figure A-12). This leads to a higher wheelset AOA, higher lateral forces (until the saturation is reached), higher L/V ratios, and increased wheel and rail wear. Furthermore, without longitudinal force, any L/V values that exceed the Nadal limit will cause wheel flange climb. Therefore, independently rotating wheels have less tolerance to any track irregularities that may suddenly increase wheel lateral forces or reduce wheel vertical forces.

The center unit of the Number 8 cars on MBTA Green Line is equipped with independently rotating wheels. As discussed in Section 4.1, combined with a low flange angle and other track and vehicle conditions, the independently rotating

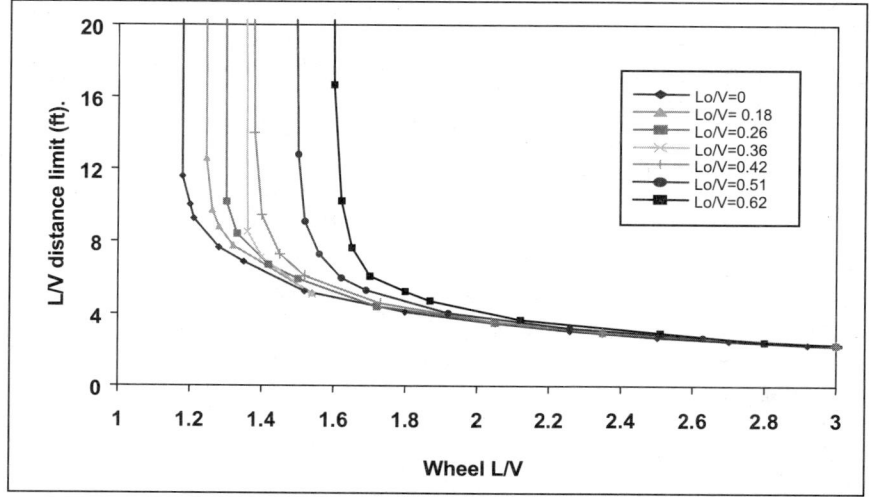

Figure A-11. Effect of wheel longitudinal force on wheel L/V ratio limit.

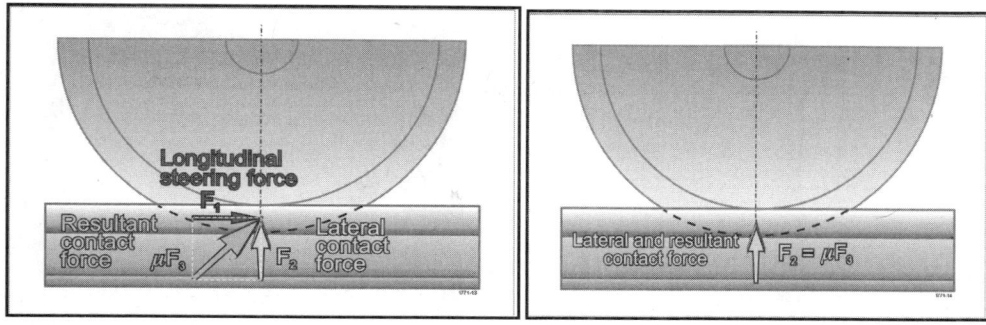

Figure A-12. Contact forces on wheels of coupled axle (left) and independent wheels (right).

wheels are more prone to derail than the wheel mounted on a solid axle on the end trucks in the same vehicle.

The center units of the low-floor cars on NJ TRANSIT are also equipped with independently rotating wheels. NJ TRANSIT reported that the independently rotating wheels have a slight tendency to climb the point of the switch if the switch is in a diverging position and is not properly adjusted. Point protections with housetops have been installed on all mainline turnouts to prevent wheel climbs. NJ TRANSIT has also observed that the wheels at the center truck have a higher wear rate compared to the wheels at end trucks due to a slight tendency to run against the low rail gage face in curving.

In summary, vehicles with independently rotating wheels need to be carefully designed to control flange climb and wheel wear. Additional control mechanisms, such as linkages or active control systems, can be used to steer wheelsets on curves and through track perturbations. Without such control mechanisms, the wheel/rail profiles and vehicle/track maintenance will need to be much more strictly controlled and monitored to prevent wheel flange climb.

4.4 CYLINDRICAL TREAD WHEELS

While most transit systems use tapered wheels, several transit systems use cylindrical tread wheels on their vehicles.

Tapered wheels have a self-centering capability. On tangent track, the primary mode of guidance is the wheel's conicity for tapered wheels. When the wheelset has a small lateral displacement from the center of the track caused by track disturbances or any asymmetry of the vehicle structure or response, one wheel will have a larger rolling radius and is moving forward faster than the other wheel. This induces a yaw motion that will move the wheelset back to the center of the track. However, if the wheelset repeatedly overshoots the center, kinematic oscillation can occur.

Cylindrical wheels tend to allow large lateral displacement even on straight track when they encounter any asymmetry in track geometry, wheel and rail profile, or other vehicle related disturbances. There is no guidance until flange contact. Wheels would run off the track without the wheel flange. Wheel flanging could be a common scenario in operation for cylindrical wheels. The wheel lateral movement is usually in a long and irregular wave, as shown in Figure A-13, based on the vehicle and track conditions. It is a different scenario from the lateral oscillation of vehicle hunting, which reflects the truck's lateral movement with constant frequency, initiated by the resonance response of the vehicle and track system.

On sharp curves and on switches and crossings, the flanges become the essential mode of guidance. Wheel flanging involves contact in a similar way for both tapered wheels and cylindrical wheels. However, the cylindrical wheels tend to produce severe two-point contact because the wheel tread tends to always be in contact with the rail and has a large difference in rolling radius compared to that of the contact point at flange. A slightly wide gage, which is a practice commonly applied in operation, can increase the RRD for tapered wheels on curves as illustrated in Figure A-14, but has no effect on cylindrical tread wheels.

On shallow curves, without flange contact, the AOA of an axle equipped with cylindrical tread wheels may be higher than that of tapered wheels due to lack of RRD. Cylindrical wheels also tend to have more flange wear resulting in thin flange due to frequent flange contact.

SEPTA uses a cylindrical tread wheel profile (63-degree flange angle) on their light rail vehicles (LRVs) operating on Routes 101 and 102. This wheel profile was inherited from previous cars operated on these lines. When new, these cylindrical wheels tend to wear quickly to a slightly hollow tread (see Figure A-15). They then stabilize in this shape for a reasonably long period. Field observations of tangent tracks on the Route 101 indicated a narrow contact band toward the gage face of the rail.

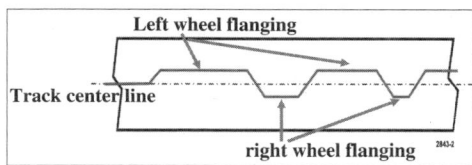

Figure A-13. Lateral movement of cylindrical tread wheelset.

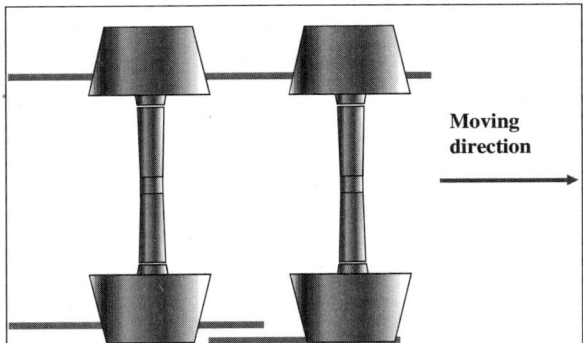

Figure A-14. Increase RRD by widening gage: (left) standard gage and (right) wider gage.

CTA uses the historical AAR narrow flange cylindrical tread profile with a flange angle of about 68 degrees. This profile was adopted by CTA in the 1930s to eliminate vehicle hunting that occurred between 60 and 80 mph on high-speed inter-city cars.

This cylindrical profile has been performing well based on CTA's report. It is likely the result of two major factors:

- CTA is a system with a high percentage of tangent tracks
- CTA has a light axle load compared to other rapid transit systems

The issue of reduced ride quality due to the lateral sliding that is usually associated with cylindrical wheels has not been raised as a problem, according to SEPTA and CTA.

4.5 WHEEL/RAIL CONTACT CONDITION ANALYSIS

The contact characteristics of a wheel and rail combination have significant effects on vehicle performance. The effective conicity resulting from the wheel tread contacting the rail head can influence the vehicle's lateral stability on tangent track, and the compatibility of the wheel flange root contacting the rail gage face can considerably affect the truck's curving performance.

In this section, the contact conditions of several wheel and rail combinations from the surveyed systems are presented to discuss their likely effects on vehicle performance. Table A-6 lists the combinations of wheel and rails. Note that all wheels and rails listed in Table A-6 are in the designed shape. They represent the initial contact situations.

Since most transit systems surveyed have no profile measurement devices except the "go" or "no-go" gages, no measured worn wheel and rail profiles were received during the survey.

Figure A-15. SEPTA cylindrical wheel wears into slightly hollow, but stable shape on Suburban Route 101, LRV.

4.5.1 Curving Indications

Figures A-16, A-18, and A-19 show the wheel and rail combinations listed in Table A-6 at the flange contact condition, which provides an indication of wheelset curving performance. Three contact patterns are observed.

TABLE A-6 Wheel and rail combinations

Wheel Profiles	Rail Profile	Rail Cant
CTA (rapid transit, cylindrical)	115RE	1:40
WMATA (rapid transit, 63 degree flange angle)	115RE	1:40
SEPTA (rapid transit, 63 degree flange angle)	115RE	1:40
MBTA (light rail, 63 degree flange angle)	115RE	1:40
MBTA (light rail, 75 degree flange angle)	115RE	1:40
SEPTA (light rail)	115RE	1:40
SEPTA (light rail, cylindrical)	100RB	1:40
New Jersey Transit (light rail)	115RE	1:40
SEPTA (commuter)	132RE	1:40

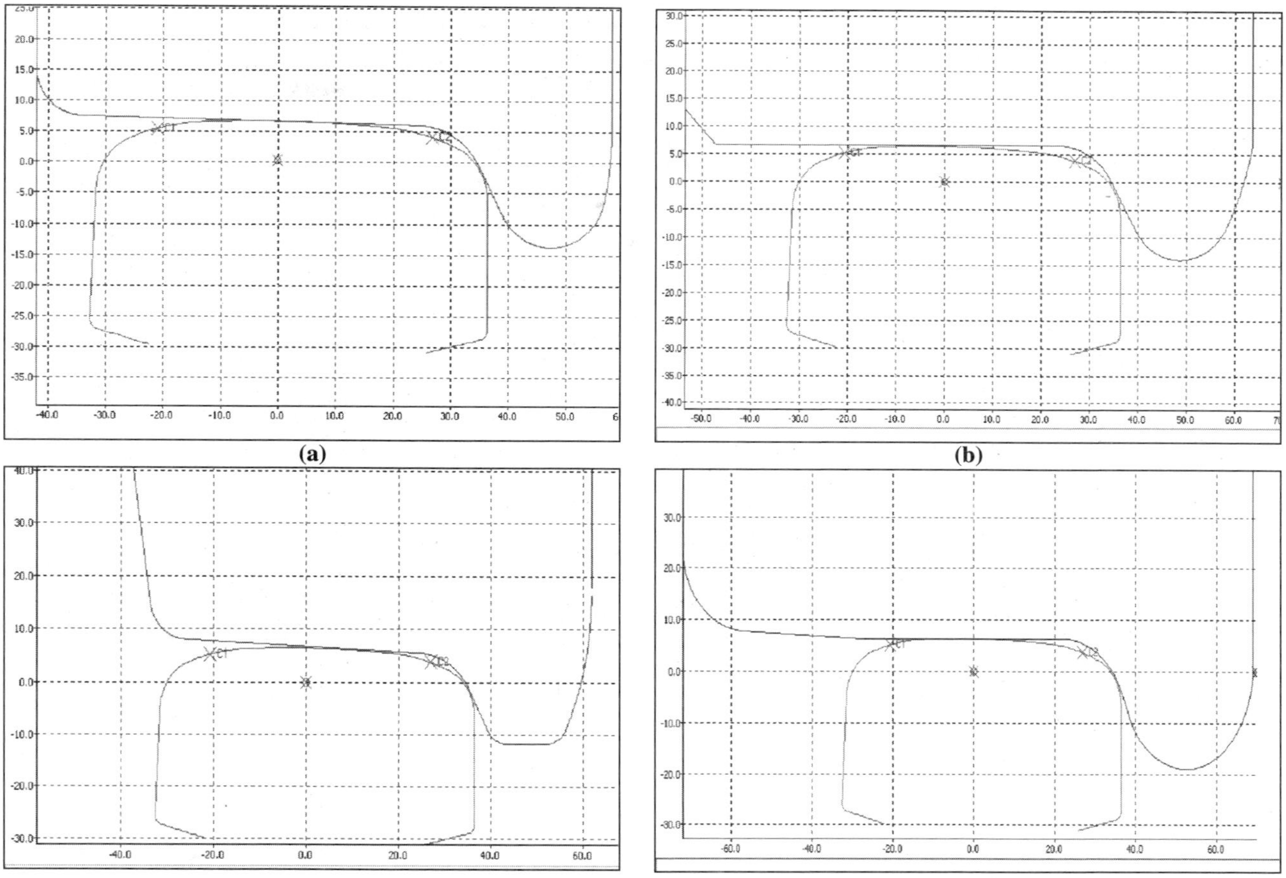

Figure A-16. Examples of severe two-point contact: (a) MBTA—light rail (Green Line, 63 degree flange angle), (b) SEPTA—light rail (101&102 Line), (c) SEPTA—light rail (Green Line), and (d) CTA—rapid transit.

Four wheel/rail pairs in Figure A-16 show severe two-point contact, which features one contact point at the wheel tread/rail head and a second at the wheel flange/rail gage face. The gap between the wheel flange root and the rail gage corner is larger than 0.08 in. (2 mm). This type of wheel and rail combination may never wear into conformal contact before the next wheel re-profiling (or rail grinding). Severe two-point contact can reduce truck steering on curves, because the longitudinal creep forces generated at the two points of contact can act in opposite directions due to the RRD at the two contact points. As illustrated in Figure A-17, the resultant steering moment would be reduced under this condition. Severe two-point contact can lead to higher wheelset AOA, higher lateral forces, higher rolling resistance, or a higher rate of wheel and rail wear.

Figure A-18 represents a medium level of two-point contact. The gap at the wheel flange throat is less than or equal to 0.04 in. (1 mm). This is more likely to wear into the conformal contact condition before the next wheel re-profiling (or rail grinding), depending on the wear rate. This type of contact starts with a similar situation as the first group with a smaller RRD between the two contact points on the same wheel.

Figure A-19 represents the conformal contact condition. Close conformal (one-point) contact provides better truck steering ability on curves than severe two-point contact.

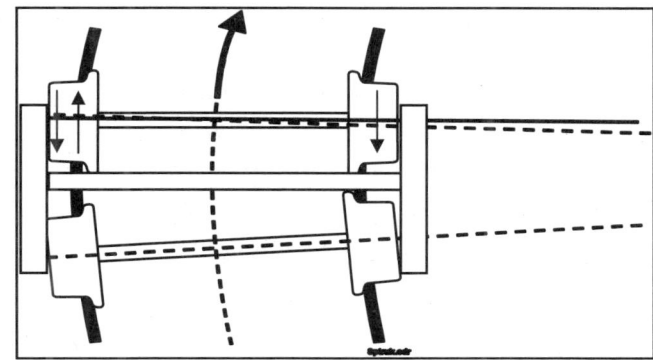

Figure A-17. Truck steering moment reduction in two-point contact, due to the opposite directions of longitudinal forces on the outer wheel.

A-18

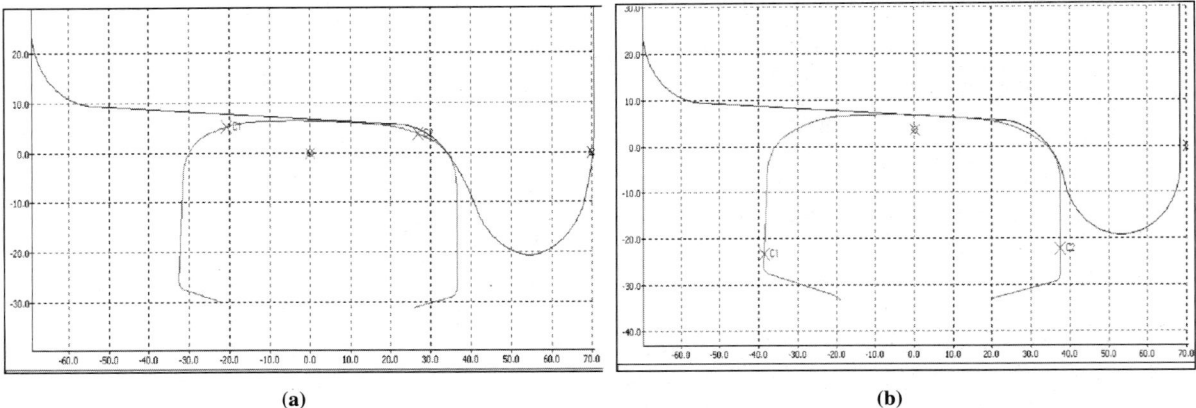

Figure A-18. Medium level of two-point contact: (a) SEPTA rapid transit and (b) SEPTA commuter.

Contact stress is also lower at the rail gage face in conformal contact because it results in a larger contact area.

Conformal contact has been a recommended contact pattern for rail operations to improve truck steering, reduce lateral forces, and reduce contact stresses to lower the risk of rolling contact fatigue (6).

4.5.2 Lateral Stability Indications

The effect of wheel and rail profiles on vehicle lateral stability on tangent track is indicated by the effective conicity. Figure A-20 illustrates contact conicity using a coned wheel. With the wheelset centered on the track, both wheels have the

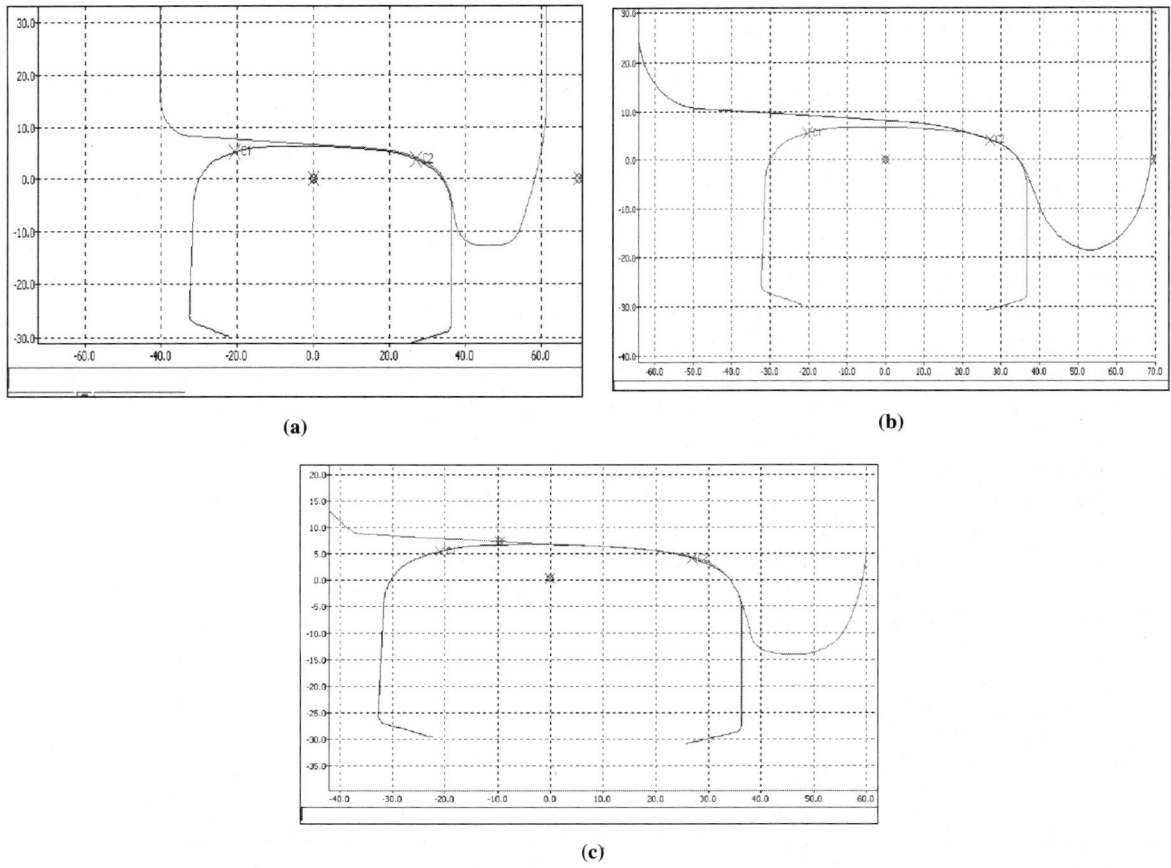

Figure A-19. Conformal contact: (a) NJ TRANSIT light rail, (b) WMATA rapid transit, and (c) MBTA light rail—interim wheel profile with 75-degree flange angle.

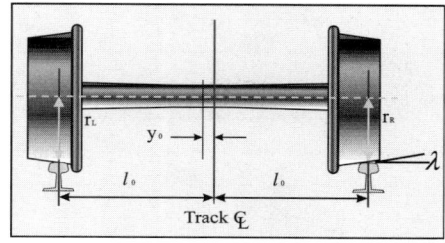

Figure A-20. Wheelset on rails, coned wheelsets (λ = conicity).

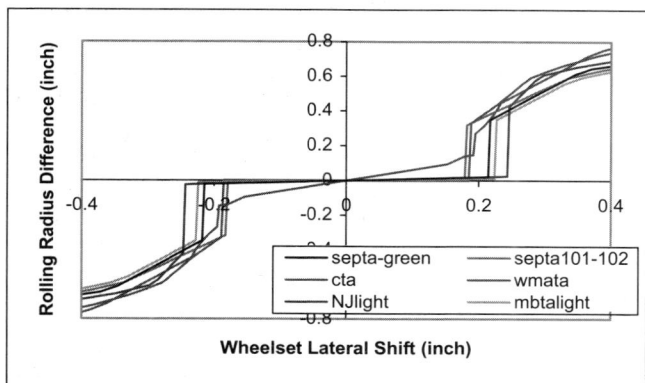

Figure A-22. RRD versus wheelset lateral shift.

same rolling radius. As the wheelset is shifted laterally (to the left in Figure A-20), the right wheel rolling radius (r_R) decreases and the left wheel radius (r_L) increases, thereby generating an RRD. The rate of change of radius for coned wheels with lateral shift depends on the cone angle λ, which is known as the conicity. In general, the effective conicity is defined by Equation A-3 (6):

$$\text{Effective Conicity} = \frac{RRD}{2y} \quad \text{(A-3)}$$

where y is wheelset lateral shift.

In normal operation on tangent track, the wheelset oscillates about the track center due to any vehicle and track irregularities (Figure A-21). Because the vehicle and track are never absolutely smooth and symmetric, this self-center capability induced by the cone-shaped wheel tread maintains the truck running around the track center. However, as speed is increased, the lateral movement of the wheelset can overshoot if the conicity is high and generate large amplitude oscillations with a well-defined wavelength. The lateral movements are limited only by the contact of wheel flanges with the rail. This unstable behavior at higher speeds is referred to as "truck hunting."

Hunting predominantly occurs in empty or lightweight vehicles. The critical speed is highly dependent on the truck characteristics. As conicity increases, the critical speed of hunting onset decreases. For this reason, it is important when designing wheel and rail profiles to ensure that the intended operating speed for a given truck is below the critical hunting speed.

Figure A-22 displays the RRD (for a solid axle wheelset) relative to wheelset lateral shift for the six wheel/rail combinations included in Table A-6. The wheelset lateral shift

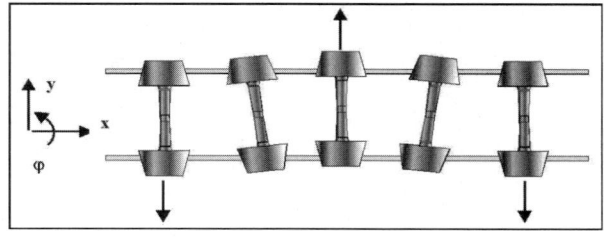

Figure A-21. Wheelset lateral oscillations on tangent track.

range before reaching flange contact is between 0.18 in. and 0.24 in. for these combinations on a standard gage of 56.5 in. Half of the slope of each line (RRD/lateral shift as defined by Equation A-3) in the lateral shift range before reaching flange contact is the effective conicity for each individual combination.

Except for the wheel used in WMATA, the combinations produced very low conicities (below 0.05) around the −0.2- to 0.2-in. lateral shift range, which indicates a low risk of lateral instability on tangent track from the aspect of wheel and rail profiles. Note that two of these wheels have a cylindrical tread.

The WMATA wheel produces a relatively higher conicity of 0.33 on average in the lateral shift range of −0.2 to 0.2 in. This value is considered to be higher than the general practice (less than 0.2) and could increase the risk of hunting on tangent track at certain speeds. However, because the truck suspensions also play an important role in hunting, a comprehensive investigation including both vehicle and track conditions at WMATA would be needed in order to determine the hunting onset speed and to conclude whether a change in the wheel profile would be required.

During the survey, WMATA reported that infrequent and somewhat transient hunting has been noticed to occur at specific track locations on the system. The lateral movements have never been severe and seem to be driven by specific combinations of vehicle and track. WMATA also makes the track gage 1/4 in. tighter on mainline tangent and on curves of less than 4 degrees. A maximum speed of 75 mph on the system is possible, but now it is restricted to 59 mph for energy conservation and equipment longevity. Therefore, it is possible that the critical hunting speed is above the current operating speed.

Conversely, the WMATA wheel profile should have better performance on shallow curves due to its large RRD compared to the other wheel profile evaluations in this study. On shallow curves (below 2 or 3 degrees), the wheelset lateral shift tends to be small and generally without flange contact. Cylindrical wheels and wheels with very low conicity can produce zero or very small RRD, as discussed above. Low RRD has a negative effect on vehicle curving performance.

4.5.3 Summary of Wheel/Rail Contact Analysis

Wheel and rail profiles are critical to system performance. In straight track, lower conicity increases the critical hunting speed. In contrast, in curved track, higher conicity enables wheelsets to achieve a lateral position near the free rolling position at small values of lateral shift. These two objectives should be achieved by controlling both wheel and rail profiles. A thorough study should be conducted before introducing any new profiles into service. Static analysis can be used as a first step in the design of appropriate profiles. Dynamic analysis is needed to verify that the designed profiles will perform well under given vehicle and track conditions. Limited track tests should also be conducted, if possible, to confirm the analysis results.

4.6 RRD, TRACK GAGE, AND RESTRAINING RAILS

To properly negotiate curves, the outer wheel needs to travel a longer distance than the inner wheel. For tapered tread wheelsets, this is achieved by running with different rolling radii (Figure A-23). The RRD required for pure rolling is a function of wheel diameter, rail gage, and track curvature. Figure A-24 displays examples of RRD for three nominal diameters of wheels on a standard track gage of 56.5 in.

Due to the limit of wheel flange/rail gage clearance (8 to 10 mm), flange contact usually occurs on curves above 3 degrees. For cars with softer primary suspensions, flange contact may happen on curves 1 or 2 degrees higher. Without severe two-point contact, the flange contact will produce a large RRD to assist wheelset curving. Figure A-14 illustrates how widening the gage, a practice that has been regu-

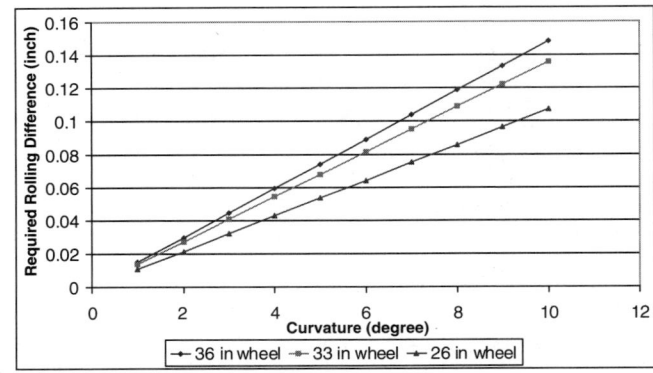

Figure A-24. RRD required for pure rolling.

larly applied in transit track maintenance on curves, can increase RRD for tapered wheels. However, the limit of gage widening needs to be set carefully based on the total wheel width, gage, and track curvature.

On sharp curves, to reduce wear at the gage face of the high rail and to reduce the risk of flange climb, restraining/guard rails are usually installed inside the low rail (Figure A-25). When properly set, the clearance between the restraining/guard rail and the low rail can allow the wheel to generate sufficient RRD to mitigate some high rail wear and transfer some lateral force to the restraining/guard rail. This clearance needs to be carefully set. If this clearance is set too tight, the RRD required for curving can be significantly reduced because the restraining rail limits wheelset lateral shift. Consequently, rolling resistance can increase considerably due to high creepages (possibly wheel slide) leading to high creep forces and wheel/rail wear.

4.7 WHEEL SLIDES AND WHEEL FLATS

All transit systems surveyed have experienced wheel sliding and, consequently, wheel flat problems. Significant maintenance efforts and cost have been devoted to reduce wheel slide and wheel flats.

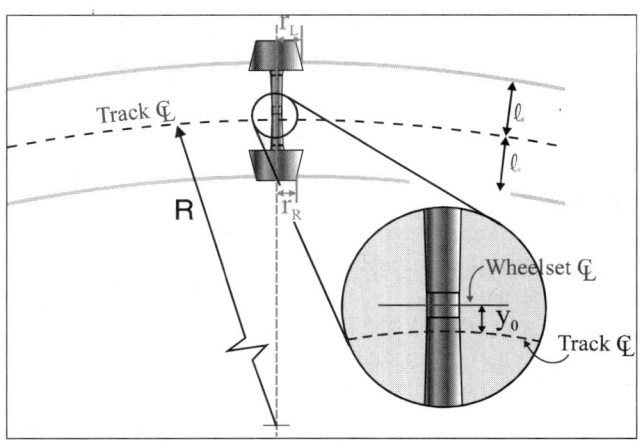

Figure A-23. RRD on curves.

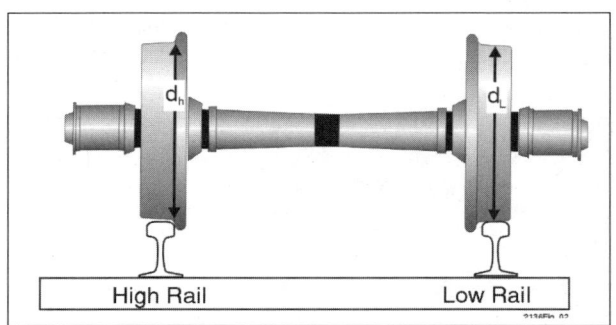

Figure A-25. Constraint of wheelset lateral shift from restraining/guard rail.

Wheel slide is caused by velocity differences between wheel and rail. It will lead to wheel flats, increased noise, significant impact forces that can damage track, and degraded ride quality. Wheel slide and flats are especially problematic during the fall season due to leaf residue contaminating the rails. Track routes closer to vegetation are most susceptible, whereas routes away from trees (elevated, subway, and between expressway lanes) are less affected.

Both traction and braking may lead to wheel slide. However, the existing literature and the survey interviews indicate that slides due to braking are more common. Magel and Kalousek report that skid flats for transit and passenger operations are due primarily to rapid and frequent brake applications under light axle loads, highly variable friction coefficients, and general over-capacity of the braking systems (7).

Based on the investigation conducted by Kumar and reported in TRB's *Research Results Digest 17* (8), chemical analyses were performed on contaminants from three U.S. transit systems indicating rust (iron oxides), dirt (silica and aluminum), and road salts (potassium, calcium, sodium, chlorine, and sulphur). Oils included petroleum products and vegetable oils as found in pine and cedar trees. The dry contaminants themselves were not problematic, but only small amounts of water or oil were needed to form pastes that significantly reduced adhesion. Heavy rains were less of a factor because they tend to wash the rails.

The flats may occur immediately as a result of abrasion (especially with cast iron tread brakes) or after additional cyclic loading (mileage) due to Martensite formation (9). Initially, the flats will have definite corners, which will round off after several miles of operation. With a deep, freshly slid flat, a concrete tie can be cracked at 56 mph. After rounding, however, a speed of 90 mph was required to crack the concrete tie (10).

The impacts caused by a single wheel flat or by a smooth wheel on special trackwork can produce transient sound pressure levels that are 7 to 10 dBA higher than the operation without the impacts (11). Therefore, the potential damage or noise will increase with the size of the impact. Note that lesser dynamic forces emanating from rough track, not complete discontinuities, can also create excessive reaction forces and noise. The noise created by impacts is usually quite noticeable above other sound and is a potential public relations problem.

The dynamic forces created by wheel impacts or rough track may be damaging to the wheel or rails themselves. Tunna reports that rail strain can easily double due to a wheel flat (10). Alternatively, the force levels may be acceptable for the wheel/rail interface but could cause problems when transmitted to secondary structures.

Wheel impacts can damage the ties, plates, ballast, and supporting or nearby structures (e.g., bridges). Even if these secondary structures are not damaged, they can become the prime radiator of secondary noise or vibration, including rattling of ties on plates and amplification by girder natural frequencies. In these cases, resilient rail fixation may reduce noise radiated from steel elevated structures and girders (11).

To control wheel slide and wheel flats, several techniques have been applied to migrate the problems, including the following:

- Pressurized spray rail cleaners.
- Hi-rail based wire brushes.
- Sander operations.
- Grit-filled gels (sandite).

Both NJ TRANSIT and SEPTA use high-pressure washers. According to an NJ TRANSIT press release (12), New Jersey invested $420,000 in a device that sprays 17 gallons per minute at 20,000 psi spray using two 250-horsepower engines on a flatcar. In addition to spray cleaning, SEPTA also operates a gel/grit delivery system and manually placed compressed sand disks ("torpedoes") on their system in periods of severe weather.

The Newark City Subway had previously tried a modified rail grinder to wire brush their rails, but the results were not satisfactory. Kumar reports similar ineffectiveness. However, Metra regularly uses an engine-powered brush on its Electric District. Metra reports acceptable cleaning results. Metra also operates additional locomotives using sanders to clean the rails during severe weather conditions.

Overseas use of special wheel-cleaning composite blocks was also noted. These are mounted on-board to clean tractive wheels and have led to higher train speeds, reduced wheel flats, and reduced train noise (8).

Prevention is the most effective measure to reduce wheel slide. Training of operators is paramount; excessive traction and braking efforts are to be avoided. The transit systems surveyed indicate that automated slip-slide controls have greatly improved this situation. These modulate one or more control parameters such as service braking pressures, dynamic brakes, motor torques, and sanding. After employing such devices, wheel flats may be reduced by roughly 50 percent under normal conditions (11). This would translate into a 50 percent reduction in periodic wheel re-profiling costs. The technology and control systems need to be further improved or developed to minimize the wheel slide and wheel flat problem.

4.8 NOISE

Every system but one expressed the recurring need for noise mitigation. Generally, noise problems fall into three categories: wheel screech/squeal, wheel impacts, and train roar. Wheel screech is usually caused by stick/slip oscillations and transmitted via the wheel plate to the surrounding air. Wheel impacts can be caused by wheel flats on smooth track or round wheels on special trackwork (e.g., crossing diamonds, switches). What is commonly known under the umbrella of "train roar" is typically wheelsets bouncing on rails due to out-of-roundness or corrugated rails because of the difference (for most materials) between a higher static coefficient of friction and lower dynamic coefficient of friction (*13*). Any technique to reduce this difference can be beneficial, such as lowering interface friction overall (lubrication) or employing a friction modifier that reverses the typical (higher) static and (lower) dynamic coefficient relationship.

In addition, since a wheelset that can generate the necessary RRD (via lateral shift) for a given curve will not require slip (creepage), wheel/rail combinations with good curving performance can reduce curving noise. Since lateral shift is affected by balance speeds, track gage, and guardrail clearances (if present), all these factors can promote or impede noise creation. Note that on the very tightest curves (less than 100-foot radius) seen during the survey, no practical wheel profile could generate the necessary RRD. In these cases, the wheels will microscopically deform at the contact points, but only up to a point. The wheels will then oscillate between stick and slip conditions, as illustrated in Figure A-26. This is analogous to a dry-friction instability, and the wheel will in turn radiate offending noise at one or more of its bending natural frequencies. Lubrication could be the key solution to noise reduction on very sharp curves.

Regarding immediate steps for mitigation of noise, the following solutions have been implemented at the various transit systems visited:

- Maintaining wheels as round as possible. This includes both removal of flats and maintenance of roundness. Noise from wheel flats tends to be directly proportional to the flat size and increases with operating speed.
- Lower creep forces via lower vehicle weights, wheel/rail lubrication or friction modification, or better curve negotiation.
- Use of resilient rail fasteners. This is most effective when the actual offending noise source is not the wheels or rails but a secondary structure (e.g., girders) responding to the dynamic forces transmitted out of the rails.
- Use of resilient wheels or of ring dampers on solid wheels.
- Adding sound absorption to stations and tunnels via surface treatments. This is not very effective at lowering noise levels, especially with ballasted track, which is already somewhat absorptive. Transit systems in general have reported mixed results after surface treatments. Such techniques should be carefully considered.

Other elements that can reduce curve squeal discussed during the onsite surveys include resilient wheels, ring damped wheels, wheel taper, primary suspension stiffness, lubrication, restraining rails, gage widening, and curvature. Appendixes A-1 to A-5 discuss these topics.

4.9 FRICTION MANAGEMENT AND LUBRICATION

Friction plays an important role in wheel/rail interface. It affects many wheel/rail interaction scenarios:

- Wheelset steering
- Truck hunting
- Wheel/rail wear
- Rolling resistance
- Traction
- Braking
- Wheel flats
- Wheel-climb derailment
- Rolling contact fatigue
- Noise

The following are major benefits of applying a friction modifier or lubrication at the wheel/rail interface for transit systems:

- Reduce wheel/rail wear.
- Reduce wheel-climb derailment.
- Reduce noise of wheel squeal.

All transit systems surveyed have applied different types of wheel/rail lubrication techniques at some level. In general, commuter operations usually apply wayside lubricators on curves, especially sharp curves. Due to frequent braking and acceleration when approaching stations, the rapid transit and light rail operators have been more cautious when applying lubrication. Table A-7 lists the lubrication practice used by the transit systems surveyed.

The Newark City Subway (light rail) has installed both wayside flange lubricators and top-of-rail friction modifiers. A relatively precise control of the amount of lubricant applied has been achieved based on the counts of wheel

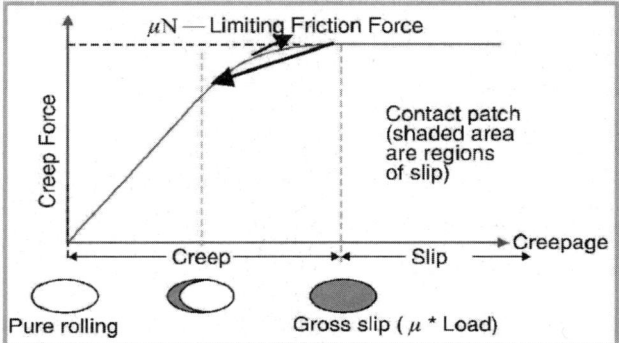

Figure A-26. Stick-slide occurred around the saturate region.

TABLE A-7 Track lubrication practice (the blank indicates no such service in that system)

System	Light Rail	Rapid Transit	Commuter
MBTA	Limited wayside lubricators for back-of-flange and restraining rail. Others have been tested	No information was received	No information was received
NJ TRANSIT	Wayside flange lubricators, top-of-rail friction modifiers, and an onboard lubrication system are under testing.		Wayside lubricators on curves.
SEPTA	Manually greased at very sharp curves	Manually greased at very sharp curves.	Wayside flange lubricators on curves over 3 degrees.
WMATA		No lubrication on mainline track. Some wayside flange lubricators are used in yards.	
Chicago Metra			Wayside flange lubricators on curves.
CTA		Wayside flange lubricators on curves with radii less than 500 ft. A few trial wayside top-of-rail friction modifiers.	

passes. NJ TRANSIT has reported successful noise reduction on sharp curves in the street and inside tunnels.

Wheel sliding that could be induced by improper rail lubrication is a major concern for some transit systems, especially in city operations. Some of them have been limiting the application of any types of rail lubrication. Therefore, to receive the benefit from rail lubrication yet avoid the negative effects, a highly structured friction management program is needed. The lubrication techniques need to be carefully selected and thoroughly tested to meet specific requirements. The ideal friction management for transit operations is to maintain a minimum level of lubrication to assure a smooth surface for the rail gage corner in curves, while avoiding over-lubrication, particularly at locations approaching stations and street crossings.

Section 3.4 of the International Heavy Haul Association's *Guidelines to Best Practices for Heavy Haul Railway Operations: Wheel and Rail Interface Issues* (6) describes different lubrication devices and techniques in detail. It also provides recommendations on lubrication practice. Although this book is focused on freight operations, some concepts in the lubrication section can be applied to transit operations. Transit operators may also need to establish guidelines that take into consideration the special features of their systems.

4.10 WHEEL/RAIL WEAR

Reducing wheel/rail wear and extending wheel/rail life is one goal that every system is making an effort to achieve.

Wear cannot be totally avoided at the wheel/rail interface due to the steel-to-steel contact of wheel and rail under heavy load. However, wear can be reduced if the contact conditions are properly controlled. Wheel/rail wear is proportional to the energy dissipated at the wheel/rail interface. Therefore, the dissipated energy can be used to define a Wear Index. This is calculated from values of creep (γ_x: longitudinal, γ_y: lateral, and ω_z: spin) and tangential force (T_x: longitudinal, T_y: lateral, and M_z: spin moment) in the contact patch for all wheels in a vehicle (Equation A-4).

$$Wear\ Index = \sum_n (T_x\gamma_x + T_y\gamma_y + M_z\omega_z) \quad (A\text{-}4)$$

The higher the Wear Index number, the greater the amount of wheel and rail wear. Equation A-4 indicates that wear increases with increased tangential forces, with increased creepages, or both. Figure A-26 shows that there are no creep forces (tangential force) or creepage in the pure rolling condition. The creep force increases as creepage increases until it achieves saturation. The creepage can continue to increase under the saturation condition. However, the resultant creep force remains the same.

As discussed in the previous sections, poor wheelset steering on curves caused by improper wheel/rail profiles such as two-point contact, low RRD, cylindrical tread wheels, or independently rotating wheels can lead to higher creepages and higher creep forces and, hence, higher levels of wear.

In addition to safety concerns, vehicle hunting on tangent track can produce much higher creep forces and creepages during wheelset yaw motion than during normal operation.

Wheel sliding generates a saturated level of creep force and very high values of creepage to cause a very high level of wear in a short distance and the formation of wheel flats. The flats are often only a few spots on the wheel surface. However, in order to remove the flats, a considerable amount of material needs to be removed around the wheel diameter during wheel re-profiling. Wheel life can be significantly reduced if flats occur frequently. Generally, a wheel can only be re-profiled three to five times before reaching the thin rim limit.

High contact stress, usually due to a small contact area, combined with high tangential forces can lead to a very high wear rate from plastic deformation of the material. High contact stress can also contribute to rolling contact fatigue (RCF). However, none of the systems surveyed indicated much concern with RCF.

High friction at the truck center bearing can produce a high turning moment that resists truck steering on curves. Under normal vehicle and track conditions, the forces generated at the wheel/rail interface can overcome the resistance at the truck center bearing, but the higher force values will be associated with higher creepages and wear rate when truck center bearing friction is high (14).

The design of trucks can also affect wear. Trucks with soft primary suspensions that allow the axles to steer in curves will generally have lower wheel-wear rates than trucks with stiff primary suspensions. However, if the primary suspension is too soft, high-speed stability may be adversely affected.

Further, any track irregularities can increase creep forces and creepages at the wheel/rail interface leading to higher wear.

The following activities can reduce wheel/rail wear:

- Optimization of wheel/rail profiles to improve vehicle curving and lateral stability and to reduce contact stress.
- Reduction of friction at the truck center bearing.
- Optimization of truck primary suspension to improve axle steering.
- Improvement of track maintenance to reduce geometry irregularities.
- Application of proper lubrication at the wheel/rail interface.

4.11 PROFILE MEASUREMENT AND DOCUMENTATION

Most transit systems use go and no-go gages to measure wheels and rails for making maintenance decisions. Only a few systems possess profile contour measurement devices. Without profile contour measurements, operation and maintenance staff may not be aware of actual wheel/rail contact conditions. The following issues related to wheel/rail profiles have generally not been addressed and documented by the systems that were surveyed:

- Initial contact conditions (new wheel contacting with new rail):
  - What is the initial contact pattern (severe two-point contact or conformal contact)?
  - What are the reasons that the specific wheel/rail profiles were adopted?
  - What are the contact characteristics of new wheel and new rail?
  - Were any simulations or track tests performed to confirm the positive performance of selected profiles?
- Stable contact conditions (stable worn shapes of wheel and rail):
  - What are the shapes of stable worn wheels and rails?
  - How long does it take for the new wheel/rail to wear into the stable worn shapes?
  - How long does the stable worn shape last?
  - What are the contact characteristics of stable worn wheel and rail?
- Contact condition of new trued wheels:
  - What are the contact conditions of new trued wheel contacting with stable worn rail?
  - What are the contact conditions of new trued wheel contacting with new rail if the trued wheel profile is different from the new wheel profile?
- Contact condition of ground rail:
  - What are the contact conditions of ground rail contacting with stable worn wheels?
  - What are the contact conditions of ground rail contacting with new wheel if the ground profile is different from the new rail profile?
- Critical contact conditions:
  - What are the worn wheel and rail profiles that may lead to flange climb or lateral instability?
  - What are the critical contact conditions in the system?

Observations in the survey indicate that there is a need to produce improved guidelines for transit operations documenting the wheel/rail profiles and contact situations under different operating conditions.

Maintaining wheel/rail system stability requires a clear maintenance objective in wheel/rail interaction. No adequate maintenance objectives can be defined without profile measurements, an understanding of the actual contact conditions, and acknowledgement of wheel/rail wear progress and patterns. Practices that are only "based on experiences" are not likely to achieve a high level of effectiveness and efficiency in operation and maintenance, as compared to that based on a scientific understanding of wheel/rail profiles and their effect on system performance.

4.12 IMPROVING UNDERSTANDING OF WHEEL/RAIL PROFILE AND INTERACTION

The presentation related to wheel/rail profile and interaction that the research team made during the survey provided a fundamental understanding of wheel/rail interaction and related issues. It also provided a forum where vehicle and track maintenance staff could discuss the common problems associated with wheel/rail interaction.

A 1-day seminar was also provided by the research team to a light rail system on the same topic with more detailed descriptions. Several staff members, including both managers and engineers, stated that they sometimes observed certain scenarios on wheels or rails but did not understand the physics behind them, and sometimes they tried different ways to improve the situation based on experience, such as improving curving or reducing rail wear, but without solid scientific evidence.

Improving the understanding through regular training or seminars on the topics related to wheel and rail profiles, wheel/rail interaction, and vehicle dynamics for transit employees (in particular, managers and engineers in maintenance groups) should be one strategic step in system improvement. With better understanding of the basic concepts of vehicle/track interaction, the operation and maintenance will be performed more effectively by making better decisions and selecting proper practices.

CHAPTER 5
CONCLUSIONS

The following is a summary of common problems and concerns related to wheel/rail profiles and interaction in transit operations identified by the survey conducted at six representative transit systems:

- Adoption of low wheel flange angles can increase the risk of flange climb derailment. High flange angles above 72 degrees are strongly recommended to improve operational safety.
- Rough wheel surface finishes from wheel re-profiling can increase the risk of flange climb derailment. Final wheel surface finish improvement and lubrication could mitigate the problem considerably. Introduction of new wheel and rail profiles needs to be carefully programmed for both wheel re-profiling and rail grinding to achieve a smooth transition.
- Without adequate control mechanisms, independently rotating wheels can produce higher lateral forces and higher wheel/rail wear on curves.
- Cylindrical wheels may reduce the risk of vehicle hunting, but can have poor steering performance on curves.
- Some wheel and rail profile combinations used in transit operations were not systematically evaluated to ensure they have good performance on both tangent track and curves under given vehicle and track conditions.
- Severe two-point contact has been observed on the designed wheel/rail profile combinations at several transit operations. This type of contact tends to produce poor steering on curves, resulting in higher lateral forces and higher rates of wheel/rail wear.
- Track gage and restraining rails need to be carefully set on curves to allow sufficient RRD and to reduce some high rail wear and lateral force.
- Wheel slide and wheel flats occur on several transit systems, especially during the fall season. Although several technologies have been applied to mitigate the problem, transit operators are in need of more effective methods.
- Generally, noise related to wheels and rails is caused by wheel screech/squeal, wheel impact, and rail corrugations. Wheel/rail lubrication and optimizing wheel/rail contact could help to mitigate the noise problems.
- Wheel/rail friction management is a field that needs to be further explored. Application of wheel/rail lubrication is very limited in transit operation due to the complications related to wheel slide and wheel flats.
- Reduction of wheel/rail wear can be achieved by optimization of wheel/rail profiles, properly designed truck primary suspension, improvement of track maintenance, and application of lubrication.
- Without a wheel/rail profile measurement and documentation program, transit operators will have difficulty reaching a high level of effectiveness and efficiency in wheel/rail operation and maintenance.
- Further improvement of transit system personnel understanding of wheel/rail profiles and interaction should be one strategic step in system improvement. With better understanding in their basic concepts, vehicle/track operations and maintenance will be performed more effectively.

REFERENCES

1. *APTA Membership Directory,* American Public Transportation Association, Washington, D.C., 2002.
2. Wu, H., and J. Elkins, *Investigation of Wheel Flange Climb Derailment Criteria,* AAR Report R-931, Association of American Railroads, Washington, D.C., July 1999.
3. Nadal, M. J., "Locomotives a Vapeur, Collection Encyclopedie Scientifique," *Bibloteque de Mechnique Appliquee et Genie,* Vol. 186, Paris, France, no date.
4. Parsons Brinckerhoff Quade & Douglas, Inc., *TCRP Report 57: Track Design Handbook for Light Rail Transit,* TRB, National Research Council, Washington, D.C., 2000.
5. *APTA Passenger Rail Safety Standard Task Force Technical Bulletin,* 1998-1, Part 1, American Public Transportation Association.
6. *Guidelines to Best Practices for Heavy Haul Railway Operations: Wheel and Rail Interface Issues,* International Heavy Haul Association, May 2001.
7. Magel, E., and J. Kalousek, "Martensite and Contact Fatigue Initiated Wheel Defects," *Proceedings, 12th International Wheelset Congress,* Qingdao, China, pp. 100-111, September 1998.
8. *TCRP Research Results Digest 17: Improved Methods for Increasing Wheel/Rail Adhesion in the Presence of Natural Contaminants,* TRB, National Research Council, Washington, D.C., May 1997.
9. Sun, J., et al., "Progress in the Reduction of Wheel Palling," *Proceedings, 12th International Wheelset Congress,* Qingdao, China, pp. 18-29, September 1998.
10. Tunna, J. M., "Wheel/Rail Forces Due to Wheel Irregularities," *Proceedings,* 9th International Wheelset Congress, Montreal, Quebec, paper 5-1, 1988.
11. Nelson, J., *TCRP Report 23: Wheel/Rail Noise Control Manual,* TRB, National Research Council, Washington, D.C., 1997.
12. "NJ Transit Unveils Aqua Track to Prevent Wheel-Slip Conditions," New Jersey Transit press release, accessed via internet, December 2003.
13. Harris, C., and A. Piersol, *Harris' Shock and Vibration Handbook,* 5th Edition, McGraw-Hill, pp. 5.19-5.20, and 40.6, 2002.
14. Wu, H., J. Robeda, and T. Guins, "Truck Center Plate Lubrication Practice Study and Recommendations," Association of American Railroads, to be published.

BIBLIOGRAPHY

Brickle, B., *TCRP Research Results Digest 26: Rail Corrugation Mitigation in Transit,* Transportation Research Board, National Research Council, National Academy Press, Washington, D.C., June 1998.

Eadie, D., and M. Stantoro, "Railway Noise and the Effect of Top of Rail Liquid Friction Modifiers: Changes in sound and Vibration Spectral Distributions," 6th International Conference on Contact Mechanics and Wear of Rail/Wheel Systems, Gothenburg, Sweden, June 2003.

Leary, "Railcar Train Track Dynamics Testing," AAR Report P-93-109, Association of American Railroads Transportation Test Center, Pueblo, Colorado, March 1993.

Shabana, A., K. Zaazaa, and J. Escalona, "Modeling Two-Point Wheel/Rail Contacts Using Constraint and Elastic-Force Approaches," *Proceedings*, IMECE2002 ASME International Mechanical Engineering Congress and Exposition, New Orleans, Louisiana, November 17-22, 2002.

Booz Allen & Hamilton, Inc., *TCRP Report 2: Applicability of Low-Floor Light Rail Vehicles in North America,* TRB, National Research Council, Washington, D.C., 1993.

Wickens, A.H., "Fundamentals of Rail Vehicle Dynamics, Guidance and Stability," Swets and Zeitlinger, 2003.

Wilson, Ihrig, & Associates, Inc., *TCRP Report 67: Wheel and Rail Vibration Absorber Testing and Demonstration,* TRB, National Research Council, Washington, D.C., 2001.

APPENDIX A1-1
MBTA

The MBTA service district includes 175 cities and towns. MBTA operates 3 rapid transit lines, 13 commuter rail routes, and 5 light rail routes (Central Subway/Green Line).

The information presented here relates primarily to light rail performance on MBTA's Green Line. The Green Line is the nation's first subway. Service under Boston Common between Park Street and Boylston was inaugurated in 1897.

Two types of light rail vehicles are currently operating on the Green Line: Number 7 cars from Japanese manufacturer Kinki Sharyo and Number 8 cars from the Italian manufacturer Breda (Figure A1-1). Both types of cars are articulated at the center truck. Number 8 cars feature low floor platform level boarding for handicapped accessibility. Although deliveries began in 1998, MBTA's acceptance of Number 8 cars remains incomplete due to various factors relating to the low floor technology's sensitivity to the Green Line infrastructure.

The information received on other MBTA lines is limited; it is discussed herein if it related to the subjects of this report.

A1.1 WHEEL AND RAIL PROFILES

By far, the most active wheel and rail profile discussions at MBTA relate to the Green Line.

The wheel profile for both the existing Number 7 cars and the new Number 8 cars originally had a flange angle of 63 degrees. Combined with other vehicle and track factors, car performance in recent years has been a priority concern on the higher speed Riverside extension of the Green Line. These factors include fairly large lateral car body motions on the existing Number 7 cars and flange climb derailment on the Number 8 cars.

The "low floor" center body of the Number 8 car is equipped with independently rotating wheels. Due to the absence of longitudinal forces in curving (as with any independent rotating wheels), the Number 8 mid-car axles produce no steering moment for vehicle curving, which has resulted in large wheel AOAs. Combined with the low flange angle, the condition presents a risk of flange climb derailment. Prompted by a series of Number 8 car derailments, MBTA undertook a thorough review of the Green Line infrastructure and reevaluated both track and vehicle maintenance practices.

At the time of this survey (late 2003), performance on the Green Line, including most notably elimination of the incidence of derailment of the Number 8 cars, had improved due to the change implemented in car and track maintenance. MBTA's remedial actions included the following:

- Where possible, rail profile grinding to remove the gage face lip and "serpentine rail wear" attributed to lateral motion exhibited by the Number 7 cars

Number 7—Two Section LRV

Number 8—Three Section LRV

Figure A1-1. Number 7 and Number 8 type car.

- Closer attention to short wavelength track geometry perturbations (previously neglected due to 62-foot mid-chord offset criteria)
- Implementation of an Interim Wheel Profile (IWP) with a flange angle of 75 degrees for the Number 8 cars

The transition from the old wheel profile to the interim wheel profile with a higher flange angle was not without complications. Due to the limited capacity of wheel re-profiling and rail grinding, many Number 7 cars, which make up the majority of the Green Line car population, were still equipped with wheels with 63-degree flange angles (or slightly higher at worn condition). Some sections of worn gage face rail were still in use (with a gage angle of 63 degrees or slightly higher). These low angles on wheel flanges and rail gage faces continued to resist the profile transition and cause fast wear of the few cars (10 to 20 cars) with 75-degree-flange-angle wheels. This high wear rate required very frequent re-profiling of these wheels (as low as 3,000 operating miles between re-profiling) in order to maintain the desired 75-degree flange angle.

Further complications related to the wheel re-profiling machine installed at the Riverside shop, which is a milling-head type. Quick dulling of the wheel cutters required the cutter head to be re-indexed after re-profiling three cars. Re-indexing of the many cutting faces required 1 to 2 days of labor. This limited the shop capacity and further slowed down the transition to a higher and operationally stable wheel flange angle.

Finally, the Green Line wheel tread design taper was 1:40, but recent worn-wheel profiles showed that the wheels consistently wore into a 1:20 taper. Thus, the IWP started with a more stable 1:20 tread taper to improve curving performance.

MBTA's current standard for rail profile installation is 115RE.

A1.2 WHEEL LIFE AND WHEEL RE-PROFILING

Previously, without a wheel re-profiling program for the Number 7 cars, wheels were condemned when they reached the thin flange limit. The average wheel life was about 18 months, or approximately 75,000 to 80,000 operating miles. During the IWP transition of wheel flange angle, wheels are being re-profiled every 2,000 to 3,000 mi to maintain the desired flange angle. The wheel can only be re-profiled three to five times before reached to the thin rim limit. Thus, the average life for these wheels was about 10,000 to 12,000 operating miles.

Elsewhere at MBTA, wheel re-profiling is performed only based on wheel flat and thin flange criteria.

A1.3 RAIL LIFE AND RAIL GRINDING

At the MBTA, rapid rail wear on the tightest curves (e.g., the 75-foot radius, ~76 degree, Bowdoin Loop on the Blue Line) allowed only 3 to 5 years between rail replacements. On such tight curvature, a practical wheelset cannot achieve the necessary RRD that would allow curving without excessive wear. To reduce flange climb risk and reduce wear, restraining rails were installed on any curve with a radius less than 1,000 ft (~5.7 degrees). The restraining rail may further reduce the desired wheelset lateral shift that is required to produce RRD. Applying lubrication at the wheel/rail interface is a potential solution for reducing wear on tight curves.

To accommodate the sensitivities of the Number 8 car and improve the Green Line track conditions in general, MBTA expended significant effort locating a grinding contractor that could remove the gage face lips on the LRV track. Only one vendor was able to rotate its stones to the necessary 75-degree angle. MBTA is gradually removing the lips where possible. However, the lack of stone clearance available in guarded or girder rail means that it cannot be ground. These gage face conditions cannot be addressed without rail replacement. Therefore, the gage face lip will not be eliminated fully on the line for at least several years.

The system does not have a programmed grinding effort for rail crown maintenance at this time. Limited rail grinding is performed only for solving immediate problems. For example, near Malden (Orange Line), noise and vibration complaints have been related to corrugations and resulting train roar. As a result, MBTA periodically grinds that section of rail to reduce train noise.

A1.4 TRACK STANDARDS

Improvements in track maintenance over the past 2 years have included grinding to remove the gage face lip (except where girder or guarded rails make this impossible) and attention to short wavelength track perturbations (previously missed due to 62-foot mid-chord offset criteria). MBTA has also focused more attention on using representative weights over track to provide loaded track surface measurements. This increased attention to track geometry maintenance is intended to improve LRV stability and ride quality and to reduce the Number 8 car's higher derailment propensity.

Because some MBTA rights-of-way were originally installed to accommodate earlier modes of public transportation (e.g., horse-drawn streetcars), some locations preclude use of current track design standards. For example, in certain locations, there is no room for proper spirals between tangent and curves or between two adjacent curves.

A1.5 FIXATION METHODS

Although much track is conventional tie and ballast design, various direct-fixation track structures were also noted at MBTA. This included use of various direct-fixation fasteners on concrete roadbeds. Except for the extra costs, MBTA is pleased with the additional track isolation gained from such practices and with the longevity of the installations.

However, MBTA personnel related a few past experiences of premature concrete tie and/or direct fixation hardware deterioration. Two examples on the Red Line were mentioned: one near the Harvard stop and one south of the JFK/UMass stop. Consequently, some removal of two-block concrete ties was performed. Also, it was felt that poor wheel roundness accelerated this deterioration. Since then, better maintenance of wheel flats has improved the situation.

A1.6 LUBRICATION AND WHEEL SLIDE

Limited wayside lubricators for back-of-flange and restraining rail lubrication are in use at MBTA. Directly lubricating the rail gage face or applying rail top lubrication or friction modifiers on tight curves have been or are being tested. Concerns that the lubrication may cause wheel sliding have slowed some implementations of rail lubricators. (Potential benefits are weighed against possible negative influences on braking distance—especially near platforms.)

MBTA personnel reported that previous problems of operator-induced wheel slides have been largely eliminated with improved slip/spin control systems (e.g., automatic sanders).

A1.7 NOISE

MBTA has made considerable efforts to resolve occasional public complaints of noise and/or vibration. Various problem sources and solutions have been implemented as discussed below.

On the Red Line, noise and vibration issues have occurred from the Park Street to St. Charles stops, near Harvard, and between the Savin Hill and JFK/UMass stops. Some of these issues could be traced to wheel flats. These have occurred more frequently on this line than the others (possibly due to heavier axle loads). As a result, MBTA is trying to gain a tighter control on wheel surfaces.

The Park Street to St. Charles locations includes a subway-to-surface portal that leads the tunnel within just a few feet of residential foundations and basements. This section was reconstructed with a resilient ballast mat in the past, but further mitigation is planned using specialty fasteners. Similar retrofits have been applied between Savin Hill to JFK/UMass.

The Harvard-to-Alewife extension was built with the intent of avoiding any potential noise and vibration complaints. As such, this section has rails over concrete slabs supported on neoprene disks. This provided 10- to 20-decibel reductions in groundborne vibration. This is a more expensive infrastructure than the use of the resilient fasteners mentioned above. Since the tunnel in this location is deeper than anywhere else within MBTA (about 130 ft deep at Porter), the earth itself provides an additional noise and vibration barrier.

A1.8 MAJOR CONCERNS AND ACTIONS

The most immediate concern on the Green Line at the time of the survey was accommodating the ongoing acceptance of the low floor Number 8 cars and improving the stability of existing Number 7 cars operating on the same line. This was being addressed through multiple, simultaneous efforts:

- Increasing the maximum flange angle on the wheels.
- Removing track serpentine wear and/or rail gage shelf (via rail replacement or gage face grinding).
- Attending to shorter wavelength track geometry perturbations.
- Replacing worn truck components on Number 7 cars.
- Continuing investigation into potential design improvements to the Number 8 cars.

Public perception of noise and vibration issues were a continuing concern. As discussed above, various problem sources and solutions have been implemented, including the following:

- Closer attention to removal of wheel flats and rail corrugation.
- Resilient track mats.
- Resilient rail fasteners.
- Slab on isolator construction.

MBTA is continuing to investigate more cost-effective remedies to improve the system performance and operating safety.

APPENDIX A-2
NJ TRANSIT

NJ TRANSIT operations include three light rail transit systems and a commuter rail system. The light rail systems are the Newark City Subway, the Hudson-Bergen System, and the River Line, connecting the cities of Camden and Trenton. The commuter system rail operation includes seven lines.

This Appendix includes only information regarding the City Subway and the commuter operations. The following subsections will each address City Subway issues, followed by commuter operation topics.

The Newark City Subway is a relatively short, light rail line of about 6 mi, built mostly on a previous canal right-of-way and operated for decades using President's Car Commission (PCC) streetcars. The Hudson-Bergen line is a new light rail line. Both light rail lines now use the same type of vehicles from Kinki Sharyo (Figure A2-1), with low platform-level boarding and low floor features fully compliant with the Americans with Disabilities Act's standards for accessible design. Wheel loads on these vehicles are approximately 12,000 pounds.

Although both lines use the same general vehicle and similar resilient wheels, the wheel profiles are different due to previously existing infrastructure on the City Subway. Foresight in planning allowed the same truck, axle, and wheel plate configurations in both the City Subway and Hudson-Bergen applications. This was accomplished by specifying different tread widths and back-to-back flange locations on tires that mate similarly on the wheel plates common to the two light rail lines.

The NJ TRANSIT commuter lines feature a mix of General Electric Arrow III EMU electric cars and diesel-hauled Comet I-V cars made by Bombardier (Figure A2-2). These cars are all approximately 90 ft in length and have about a 15,250-pound wheel load.

A2.1 WHEEL AND RAIL PROFILES

A2.1.1 Light Rail

Wheel profiles on the Newark City Subway follow an ORE (European Railway Organisation) standard, which is similar to the Pittsburgh LRV profile with a peak flange con-

Figure A2-1. NJ TRANSIT city subway low floor LRV.

Comet Type Car

Arrow MU Type Car

Figure A2-2. Commuter rail cars.

tact angle of 75 degrees and a 1:20 tread taper that provides for one point of contact with the 115RE rail installed on a 1:40 canted plate. The back-to-back dimension is 54.125 in. The flange top is flattened for lower contact stresses for the flange bearing frogs installed at the special trackwork. The Hudson-Bergen line uses an AAR-1B profile (75-degree flange angle) wheel design with 53.375 in. back-to-back spacing. Wheel diameter is 26 in. for both lines.

The center truck of the LRV low-floor cars is equipped with independent rotating wheels. Due to dual articulation located outside the center truck boundaries, the wheels have a slight tendency to hard curving at the point of switch when the switch is in a diverging position and not properly adjusted. City Subway personnel have installed point protections with housetops on all mainline turnouts to prevent wheel climbs.

It has been observed that the wheels at the center truck have a higher wear rate compared to the end trucks due to a slight tendency to run against the low rail gage face in curving. The wheel wear has been monitored using a MiniProf device to predict the interval before re-profiling.

The standard Newark City Subway rail profile is 115RE, purchased in 1984 to replace 100RB.

A2.1.2 Commuter Rail

All NJ TRANSIT commuter cars operate on a common, 32-in. diameter wheel. The wheel profile is an AAR narrow flange with a 72-degree peak flange angle and 1:40 tread taper. In 1988, problems with vehicle lateral instability prompted NJ TRANSIT to replace the 1:20 taper with the 1:40 tread taper.

Also, prior to 1999, the peak wheel flange angle was specified at 68 degrees. In the 1990s, a few low-speed flange climb derailments were generally attributed to unfortunate combinations of the following contributing factors:

- Negotiating special trackwork
- Dry rail
- Newly cut wheels

As a result, NJ TRANSIT commuter operation has since adopted a higher peak angle of 72 degrees. The profile transition period required approximately 1 year before all the wheels being seen in the shop had the steeper flange angle.

NJ TRANSIT commuter operation has a MiniProf device. However, wheel profiles are generally only measured in such detail after a derailment or other serious incidents.

On the NJ TRANSIT commuter operations, the current standard rail is AREMA 136 RE for new installations. Currently, 132 RE rail is still dominant throughout the system, and weights from 105 to 155 lbs./yd may also be found.

A2.2 WHEEL LIFE AND WHEEL RE-PROFILING

A2.2.1 Light Rail

With only 2.5 years of operation on a new fleet of cars (at the time of this survey), the LRVs of the Newark City Subway had an average of only about 70,000 operating miles. Based on a MiniProf survey of every wheel, NJ TRANSIT extrapolations predict 7-year (200,000- to 250,000-mi) wheel/tire lives.

The City Subway uses a lathe type wheel re-profiling machine. The first round of re-profiling may start after the cars reach an average of 100,000 mi of operation.

A2.2.2 Commuter Rail

NJ TRANSIT commuter engineers estimate a typical wheel life of 250,000 mi. Their wheel tread allows for about four turns during the wheel's life. NJ TRANSIT commuter has two milling type re-profiling machines: one at MMC and one at Hoboken. In terms of capacity, four to eight cars can be re-profiled per shift per site.

A2.3 RAIL LIFE AND RAIL GRINDING

A2.3.1 Light Rail

Rail on the Newark City Subway was all replaced in 1984. Some tight curves have shown more wear and have required replacement on a 5- to 6-year cycle. Restraining rails are installed on any curve with a radius less than 600 ft. On some very tight curves (60 and 80 ft in radius), a lip has formed on the inside rail. An initial attempt to lessen this wear pattern (by moving the guardrail for a narrower flange way) actually promoted the lip formation. NJ TRANSIT is now planning to increase the track gage to 57 in. and allow a 2-in. flangeway clearance between the stock rail gage face and the restraining rail on curves with 82-foot and 60-foot radii. It is expected better curving performance could reduce this wear.

In several locations of 1984-era rail, some corrugations remain as created by the PCC cars operating before 2001. However, based on NJ TRANSIT evaluations, the new light rail vehicles appear to be flattening these waves. Perhaps this is related to the different traction system on the new cars, but it may also be related to different responding wavelength of the vehicles. NJ TRANSIT plans to remove the remaining corrugations with a rail profile grinding project.

A2.3.2 Commuter Rail

NJ TRANSIT commuter lines have about 900 curves with the tightest being sharper than 10 degrees (~570-foot radius).

Rail wear was mentioned as an issue for a few curves. In these cases, rail life may be as short as 6 years. Two example scenarios were discussed:

- At a grade crossing near Gladstone, the curve is elevated to 1.5 in. underbalance (designed for 45 mph traffic), but traffic actually operates around 25 mph due to the crossing. This results in faster low rail wear and some track movement.
- Some sharper curves, which would normally receive flange lubrication, have no lubricators installed due to other concerns such as losing traction or braking capability on a grade.

NJ TRANSIT commuter operations report no programmed rail profile grinding operations. Limited spot grinding is used to return the rail shape to a new profile, but it is not generally targeted toward specific rolling contact fatigue issues. Newly installed rail is commonly ground after 1 year to remove surface defects or corrugations.

Approximately twice a year, an ultrasonic and induction rail inspection is conducted across the system, uncovering 8 to 15 defects each time. Neither rail shelling nor corrugations are significant issues.

A2.4 TRACK STANDARDS

Track Geometry Standards (known as MW4) are mainly used by the commuter rail system and are used as a guide for the Newark City Subway. City Subway staff report that City Subway follows much tighter classifications for its operation than FRA would prescribe. For example, the light rail track gage dimension is a nominal 56 1/2 in., with a +1/4 in. and −0 in. tolerance, and the maximum operating speed is 50 mph. The FRA standard for commuter rail gage is a nominal of 56 1/2 in. with a +1 in. and −1/2 in. tolerance for speeds above 60 mph.

The Newark City Subway experienced a derailment during turnout negotiation on a Number 10 Samson switch. It has been determined that the AREMA 5200 detail at the Samson type switch and the quality of point adjustment to the stock rail was not adequate and may always create a hazard. Consequently, housetop point protection was retrofitted to all switches.

FRA track geometry standards apply on the NJ TRANSIT commuter lines. The National Railroad Passenger Corporation (Amtrak) track geometry car is currently used across the system quarterly. The Amtrak inspections of the 550-mi track typically yield about five to eight track geometry defects. NJ TRANSIT intends to perform 8 to 10 inspections per year. Programmed surfacing is performed on a 5- to 10-year interval and tends to follow the tie replacement cycle.

A2.5 FIXATION METHODS

The Newark City Subway currently uses wood ties on ballast, except in stations where dual block wood is set in concrete. A planned service extension will be near both an historic church and a performing arts center. Therefore, a floating slab design has been specified to reduce noise and vibration. The cost of this is estimated to be three to five times that of conventional track.

The commuter operations at NJ TRANSIT use conventional wood tie/cut spike construction with premium fasteners on some curves. That is, curves greater than 2 degrees (~2,900-foot radius) are gradually being re-fit with Pandrol fasteners and lag screws. Previously, NJ TRANSIT used hairpin-type fasteners in the curves, but the fasteners did not hold.

A2.6 LUBRICATION AND WHEEL SLIDE

A2.6.1 Light Rail

The Newark City Subway applies a variety of rail lubrication methods in its system. Wayside flange lubricators and wayside top-of-rail friction modifier systems are operating, and an onboard lubrication system is currently under test.

In the yard there are eight lubricators using regular grease for the flange side and the back of the wheel.

Wayside top-of-rail friction modifiers have been installed at the 60-foot and 82-foot curve radius turnaround loops at Penn Station (tunnel) and at the 100-foot radius (outdoor) curve at Franklin Street. Site inspections confirmed that no wheel screech was perceived at these locations. NJ TRANSIT reports no adverse effects of weather on the vehicle performance using the friction modifier outdoors.

A concern for the Subway during autumn and spring is the so-called "black rail," a slippery condition caused by falling leaves combining with morning dew and dust. When wet, the leaves are smashed by passing wheels and become a low-friction contaminant. This black rail condition can cause adhesion and braking problems systemwide. Efforts were made to improve the resulting low friction conditions via track cleaning with an electric rotating brush, but NJ TRANSIT did not report success. Rather, the brushing operation tended to merely distribute the contaminant evenly across the rail. Since then, NJ TRANSIT has procured a hi-rail water jet cleaner operating at 20,000 psi with much improved results.

A2.6.2 Commuter Rail

About 80 wayside lubricators are installed on the commuter rail system curves. There is an ongoing debate within NJ TRANSIT about the minimum curvature that should receive a wayside lubricator. A systemwide review of rail profiles and lubricator placement is underway.

As mentioned previously, some curves that would normally receive flange lubrication have no lubricators installed because of other concerns, such as losing traction or braking capability on a grade.

Leaf residue on the tracks is also a seasonal problem for the commuter operations.

A2.7 NOISE

Noise and vibration is an important issue for the Newark City Subway, especially for the rail sections that are close to residential areas. NJ TRANSIT has oriented some of their rail lubrication efforts to reduce noise levels as well as wear. At a turn-around curve (82-foot radius) in the Vehicle Base Facility yard and a few other locations, the wayside flange lubricators are used to reduce wheel squeal and wear. A similar success has been implemented at a sharp, in-street curve at Franklin Avenue via a wayside top-of-rail friction modifier. Also recently, squeal noise has been reduced underground at the sharp Penn Station curve via a top-of-rail friction modifier. As mentioned, the Newark City Subway is also making efforts to improve vehicle curving by properly adjusting rail gage and flange way clearances, which should also reduce the noise on curves somewhat.

Unlike the City Subway, noise has not been an important issue on the NJ TRANSIT commuter operations. This is expected, given that commuter systems often operate with greater separation from residential and business areas.

A2.8 MAJOR CONCERNS AND ACTIONS

A2.8.1 Light Rail

As a newly updated system overall, the Newark City Subway is maintaining a high level of operational quality with extensive efforts toward preventative maintenance. Both rail and wheel wear are being closely monitored, as evident by the use of MiniProf data from every wheel.

Various existing mitigation techniques oriented toward wear, noise, and safety have been implemented:

- Wayside flange lubrication.
- Wayside top-of-rail friction modification.
- Special trackwork point guards.
- Optimization of restraining rails on curves.

Additionally, prototype onboard flange lubrication is being tested and a program of preventative rail profile grinding is planned.

A2.8.2 Commuter Rail

NJ TRANSIT's commuter rail wheel/rail profile maintenance is an ongoing process. Daily flange width and wheel flat inspections, as well as suitable capacity in the two wheel re-profiling shops result in good maintenance of wheel tread profiles.

Past problems and solutions for the NJ TRANSIT commuter system have included the following:

- Vehicle hunting—reduced by implementing 1:40 tread tapers.
- Low-speed flange climb at special trackwork—improved by implementing higher flange angle wheels.
- Slow-speed derailments in yards (especially with newly cut wheels)—reduced by giving greater attention to yard track quality.
- Low-speed flange climb when local operations are considerably below the designed balance speed—reduced by reengineering elevations at some curves.

APPENDIX A-3
SEPTA

SEPTA has a very diverse infrastructure with operations including commuter, rapid transit, and light rail.

Both the City Transit (Green Line) and Suburban Light Rail Lines (Routes 101 and 102) use similar 50-foot long Kawasaki LRVs. However, the wheel profiles and track gage are different between the city and suburban lines.

SEPTA's three rapid transit lines are the Market-Frankford (Blue) Line, the Broad Street Subway (Orange), and the Norristown Route 100 (Purple) Line. The Blue line operates 55-foot long Bombardier M-4 stainless steel cars. The Orange Line has a fleet of Kawasaki B-IV cars each 67.5 ft in length. The Purple Line (Route 100) has a fleet of N-5 cars from Bombardier.

Regarding commuter operations, the majority of SEPTA vehicles are 85-foot long Silverliner type vehicles from Budd, St. Louis Car, and GE. Other types of commuter cars include 85-foot electric push-pull cab cars and coaches from Bombardier.

A3.1 WHEEL AND RAIL PROFILES

The diameter of all LRV wheels is 27 in. The wheel profile for the LRV cars on the Green Line has a 63-degree flange, a 1:20 tread taper, and a flat top flange that may help to reduce the contact stress as wheels pass special flange-bearing trackwork.

A cylindrical tread wheel profile is applied on the LRV cars operating on Routes 101 and 102. This wheel profile was inherited from previous cars. Analyses of tolerances for the flange root width show this to have a peak flange angle that is between 60 and 65 degrees. When new, these cylindrical wheels tend to wear quickly to a slightly hollow tread, as shown in Figure A3-1. They then stabilize to a reasonably constant shape. Field observations of tangent tracks on the Route 101 indicated a narrow contact band, skewed somewhat towards the gage face of the rail.

The light rail lines use 100RB rail. The rail gages are wider (ranged from 62.25 to 62.5 in.) than standard gage of 56.5 in.

On the Orange (Broad Street Subway) Line cars, 28-in. wheels are used with a 63-degree flange angle and a 1:20 tread taper. Except for the commuter rail lines, the new rail laid are 115RE . However, former rail standards have left 80- to 100-lb/yd rail in some sections.

Figure A3-1. SEPTA cylindrical wheel wears into slightly hollow but stable shape on Suburban Route 101, LRV.

The regional (commuter) line cars use 32-in. diameter wheels with 75-degree peak flange angle and 1:20 taper. The current rail standards for the commuter rail system are the 115RE and the 132RE.

A3.2 WHEEL LIFE AND WHEEL RE-PROFILING

The light rail lines (including the Norristown route) achieve about 150,000 to 200,000 mi between re-profiling. The average wheel life is about 10 years.

The City Transit LRV (Green Line) wheels are generally re-profiled due to flange issues. Field inspections showed that these wheels often encounter street debris in the girder rail flangeway. As such, they experience excessive riding on the top of flange. The 101/102 LRV wheels are re-profiled at about a 5-year interval, based on a predicted usage of 33,000 mi per year.

SEPTA has two re-profiling machines, a lathe type re-profiling machine and a milling-head re-profiling machine. The lathe machine has a single-point cutting tool that produces a smoother surface finish compared to that from the milling machine.

The milling machine has a cutting head with many small cutters (staggered to form the wheel profile). SEPTA expressed particular interest in any potential flange climb effects caused by smoothness differences between left and right wheels on the same axle. Such differences have been seen when using one sharper cutting head and one dull head.

The SEPTA wheel diameter tolerances after re-profiling 1/8 in. within the same axle, 1/4 in. axle-to-axle in the same truck, and 1/2 in. truck-to-truck difference in the same car.

As with all the systems visited, SEPTA has experienced low-speed derailments, and almost all of them were flange climbs in yard tracks. Some of these were associated with newly re-profiled wheels. Wheel surface roughness after wheel re-profiling, combined with SEPTA's low flange angles, could considerably reduce the L/V limit ratio required for wheel climb.

On the commuter lines, operating miles are not tracked and therefore wheel lives are not known. Generally these wheels are trued for flat spots caused by braking and/or rail contaminants.

A3.3 RAIL LIFE AND RAIL GRINDING

Rail lives vary from 6 years (tighter curves) to 40 years (tangent track). Curves over 5 degrees (~1150-foot radius) tend to wear quickly and are typically replaced within 5 to 7 years.

On the rapid transit lines, fast wear in some tangent areas can be attributed to significant use of track brakes under the cars. Alternating gage face wear between left and right rail was reported at a certain locations. The causes are still under investigation.

As shown in Figure A3-2, the rail at certain sections of the Green Line had significant wear or surface damage. The damage is likely caused by wheel impacts upon street debris in the girder rail flangeway. This can locally lift the tread contact and cause wheel impacts at an adjacent section.

For the wide gage light rail lines, SEPTA owns an 8-stone grinding machine. Rail profile grinding is targeted toward producing an 8-in. rail head crown radius.

For standard gage lines, contractors are used for infrequent grinding. SEPTA now replaces rails in tunnels when profile or surface problems are advanced. During recent experience with rail grinding in a tunnel, it was found that the spread of grinding dust, air contamination, and expensive station cleanup made rail grinding in those areas unfeasible with the available equipment.

SEPTA reported that asymmetric grinding on some curve sections successfully improved vehicle curving and resulted in reduced wear and noise.

Like other transit systems, SEPTA has had track corrugation problems at specific locations. Rail grinding is required periodically in these zones to remove severe corrugations.

A3.4 TRACK STANDARDS

Commuter rail lines follow FRA track safety standards. The light and heavy rail track is maintained to SEPTA internal track standards. In brief, SEPTA track geometry standards are similar to the FRA standards, although oriented toward 31-foot mid-chord lengths. (Gage specifications are equal to the FRA rules. Alignment specifications are under 1/2 in. of the FRA allowances. Vertical profile and cross level rules are similar to the FRA rules, and the track twist rules are slightly under the FRA allowances.)

Light rail lines are designed to 4.5 in. maximum underbalance. All other lines allow up to 3 in. underbalance. Head-hardened rails are installed on curves. Guardrails are installed for rapid transit curves wherever tighter than 750 ft in radius.

To improve vehicle curving and to reduce gage face wear of the high rail on tight curves, track gage is intentionally widened up to 1 in. in places. However, SEPTA has concerns about how much worn rail conditions effect the optimum effective gage on different curvatures.

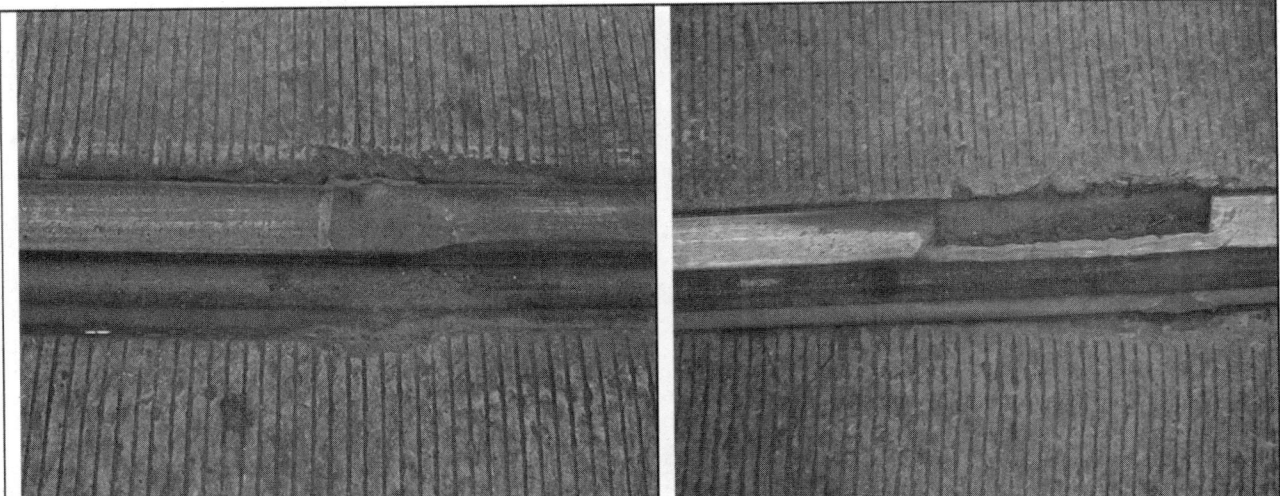

Figure A3-2. Rail surface damage on Green Line.

Regarding commuter operations, a track geometry car inspects track every third month. Walking track inspections are performed once a week for sections having less than 5 million gross ton (MGT) traffic per year and twice a week for sections having traffic more than 5 MGT per year.

Various maintenance intervals are used on the light rail and rapid transit lines (including limited cross tie and rail replacement, and surfacing where necessary). As a long-term goal, SEPTA is planning to achieve the track standards one class higher than the FRA specification.

A3.5 FIXATION METHODS

Rail fixation for the various SEPTA rapid transit lines ranges from direct fixation (e.g., wood half-ties set in concrete) to wood ties on ballast (at grade and elevated track). Light rail lines also have areas of direct pour concrete fixation. In these cases, the rails are initially held in place every 6 ft with Pandrol clips and a steel beam tie. Then, the rails are fully embedded in concrete with only gage face clearance left in the concrete. However, tracks with this installation method show that the concrete can rupture prematurely near battered joints.

For commuter track, both wood ties on ballast and booted two-block concrete ties are used.

A3.6 LUBRICATION AND WHEEL SLIDE

Rapid transit operations include up to 40-degree (150-foot radius) curves. Such curves are manually greased daily. SEPTA is hoping to improve flange grease controllability and efficiency on these curves by installing through-holes on the restraining rails, along with grease fittings and automatic pumps.

On the commuter lines, most curves over 3 degrees (~1900-foot radius) have wayside flange lubricators. The commuter lines include 12-degree (~480-foot radius) curves near a regional station. This location formerly caused excessive wheel and rail wear. Now liberal rail greasing and 15 mph speed restrictions are used to minimize wear.

Slippery rails due to leaf residue in the fall months are a major concern. In the fall, 60 to 80 percent of regional train delays are leaf-related. In 2002, this issue caused the delay of 2,357 trains. Wheel slides due to rail contamination lead to flat spots. Dynamic impacts due to these flats can damage the track, induce noise, and affect ride quality.

To mitigate the seasonal problem, SEPTA cleans the track on the Norristown line, and on regional lines during their short, 3-hour overnight work window. This is done with both advanced techniques (former locomotives now called "Gel Cars" with a 5,000-psi high-pressure washer and traction gel applicators operating at 10 mph) and more traditional methods, such as by applying sand with locomotives, and by manually placing small solid disks of compressed sand on the rail head.

A new, speed-sensing, dynamic braking control system is being implemented in the rapid transit operation to reduce wheel sliding during braking.

A3.7 NOISE

The rapid transit lines (Market-Frankford and Broad Street) and the City Transit light rail lines converge under the Philadelphia City Hall. This array of tunnels and stairwells is excessively noisy due to the combinations of squeal, flanging noise, wheel impacts, and rolling noise.

The Broad Street cars are subjectively deemed to be quite noisy, although it is believed that they were somewhat quieter when new. At least one subway station was retrofitted as a means of noise reduction. However, the noise level was not satisfactorily reduced in this station, even after the additional wheel/rail smoothing, sound absorption, and barriers. Attention to this issue continues, because the root problem source remains somewhat undefined.

The eastern portion of the Market-Frankford (Blue) line has required significant rail head and gage face grinding to remove corrugations (and associated noise) on tangents.

The City Transit (Green) line has flange-bearing wheels for special trackwork. The route has one nonflange bearing frog with level points and wings. A depressed point frog was installed at one location, but immediate public complaints resulted in replacement with a level style. Dynamic braking also tends to reduce noise by reducing wheel sliding.

A3.8 MAJOR CONCERNS AND ACTIONS

SEPTA inherited a wide array of infrastructure from preceding entities. This includes different track design standards, different vehicle types, and different track gages. SEPTA also inherited problems resulting from deferred maintenance by previous railroads. At one point, over 600 defective welds were found in track as a result of poor welding practices. This number has steadily decreased as SEPTA continues to put forth significant effort into improving track maintenance.

Similarly, the regional commuter lines were taken over in 1982 with immediate track geometry and rail condition problems, requiring several years of continuous improvement of track geometry to achieve desired quality conditions. Over the past 20 years, almost all tracks in the commuter rail system have been upgraded with continuously welded rail.

Lubrication continues to be an area of development at SEPTA. Rail wear issues primarily drive this effort. Although still employing manual track greasing in some rapid transit locations, more advanced track-based systems are slowly being implemented. The next step in this process is a trial installation of through-hole grease fittings on guardrails for flangeway lubrication.

SEPTA has experienced infrequent derailments that fall into two categories: train handling (traction, braking, excessive speed) and slow-speed flange climb (at more severe curves, yards, climbing switches, and sometimes soon after re-profiling).

Wheel sliding and the resulting wheel flats are a major issue and problem, especially during the autumn season. Several techniques have been applied to ease the problem, but the need for developing more effective technology to remove rail contaminants is needed.

The cause of rail corrugation is still under investigation. Both rail grinding and rail replacement have been conducted to remove corrugation.

APPENDIX A-4
WMATA

WMATA has a large degree of standardization. Only three car types have been used in the system's 35-year history: the original Rohr cars are still in use, as well as Breda cars purchased in the early 1980s and CAF cars recently delivered. The Rohr cars have an Atchison/Rockwell suspension with good curving performance. This type of car is deemed slightly more prone to hunting when trucks are worn. The Breda cars have a longer wheel base (5 in. longer) than Rohr cars, and a slightly stiffer primary suspension. Over time, the Breda cars have been involved in a few flange climb derailments. The cars have approximately 13,200-pound maximum wheel loads and use 28-in. diameter wheels. (The weight of the Breda is approximately 81,000 pounds.)

WMATA is a relatively new system compared to the other transit systems surveyed. Therefore, its track layout contains fewer tight curves. The tightest curve in WMATA has a radius of 250 ft (~23 degrees). An extensive preventative maintenance program results in good ride and operational qualities. However, a few recent incidents of flange climb derailments have raised concerns related to wheel profiles and track gage.

A4.1 WHEEL AND RAIL PROFILES

The original WMATA cars were supplied with cylindrical wheel profiles that resulted in excessive wheel and rail wear. To reduce the excessive wear, a field experiment was performed during 1978 and 1979 to select a wheel profile with better wear performance. The wheels in 12 trucks of three Rohr car series were machined to various wheel profiles. Wear rates were analyzed. This test led to the adoption of the "British Worn" profile as WMATA's standard wheel shape. This profile has a flange angle of 63 degrees.

In recent years, there have been several incidences of flange climb derailment—most of them at yard switches. Generally, these derailments were caused by multiple factors. Among the derailment factors, observations included newly re-profiled wheels and dry rails. Under such conditions, the friction coefficient between wheel and rail can be quite high. In combination with WMATA's low flange angle, the L/V ratio limit before precipitating a flange climb derailment can be considerably low. Also, wheels with low flange angles have less tolerance to any unexpected track irregularities.

To improve margins of safety, WMATA is considering a design wheel profile change to a higher flange angle. However, the system stability during such a transition is being carefully considered.

The current standard rail is 115RE for new installation. Head-hardened rail is installed on tight curves.

A4.2 WHEEL LIFE AND WHEEL TRUING

WMATA wheels show a typical life of about 400,000 mi, or 4.5 years of operation. Between 3 and 5 re-profiling operations are possible before reaching to the thin flange or thin rim limits. As with every system, wheel flats can be a problem during the fall season each year. The autumn leaves are a major cause of wheel re-profiling. This has periodically overloaded the wheel shops, requiring a few weeks of overtime labor to remove flats.

WMATA has two types of wheel-re-profiling machines: milling and lathe. Shop personnel report significantly smoother finishes with the lathe machine. Consequently, after the use of the milling type machine, WMATA wheels get a minimal pass as the final step of re-profiling (known as the "air cut," where no significant material is removed). Perhaps more importantly, all wheels are now manually lubricated immediately after re-profiling.

Wheel diameter tolerances after re-profiling are 1/16 in. within the axle, 1/4 in. axle-to-axle in the same truck, and 1/2 in. truck-to-truck in the same car.

WMATA believes that most wheels are trued at least once per year as a result of its "no flat" policy.

A4.3 RAIL LIFE AND RAIL GRINDING

Generally, tangent track and most low rails at curves on this system have retained the original rails. Rail replacement is performed more frequently on sharper curves. WMATA allows 1/2 in. of gage face wear. Thus, the transit system reports that curves greater than 7.5 degrees (~760-foot radius) last 3 to 5 years, and frogs last about 8 years. Rails are not generally re-laid due to the extra effort required.

An outside contractor performs grinding annually. Locations for rail grinding are specified based on subjective evaluations of ride quality and noise.

A4.4 TRACK STANDARDS

Maximum speed on the system is 75 mph, but currently it is restricted to 59 mph for energy conservation and equipment longevity.

Allowable track gage, alignment, profile, and cross level deviations tend to make up one-third to one-half of the tolerances found in FRA rules for Class 3 (60 mph) track.

Designed track gage also varies by curvature:

- 1/4 in. tight on mainline tangent to 4-degree curves.
- Standard on 4- to 16-degree curves.
- 1/2 in. wide above 16-degree curves.
- 3/4 in. wide above 16-degree curves with restrained rail.

WMATA is installing guardrails on all switches corresponding to less than a 500-foot radius (~11.5-degree). Also, curves with less than a 800-foot radius (~7.2-degree) are equipped with guardrails. Guardrail clearance is set to 1 7/8 in.

The sharpest yard curve is 250 ft in radius (~23 degrees). The sharpest mainline track curve is 755 ft in radius (~7.6 degrees). Secondary and yard tracks are designed to 4.5 in. underbalance operation. Tie plates are standard 1:40. WMATA has attributed some flange climbs on special trackwork to the lack of rail cant.

A4.5 FIXATION METHODS

All WMATA surface tracks are crosstie on ballast construction. However, both elevated and underground tracks have direct fixation via resilient (rubber/steel tie pad) fasteners. Stiff (250-kip/in.) and softer (150-kip/in.) fasteners have been tried, with the softer versions preferred.

A4.6 LUBRICATION AND WHEEL SLIDE

No lubrication is performed on mainline track, but some traditional wayside flange lubricators are used in yard tracks and guard rails. Past trials of both on-board stick lubricators and wayside top-of-rail friction modifiers proved to reduce noise but caused wheel slide and were hard to maintain. Therefore operations precluded the application of lubrication on mainline.

As with other lines, leaves are a problem in the fall causing excessive wheel flats. WMATA does not attempt to clean rails; rather operating speeds are reduced and selected safety stops are used to reduce the effects of leaves.

A4.7 MAJOR CONCERNS AND ACTIONS

Major issues for WMATA relative to wheel/rail profiles are the following:

- Improving wheel flange angle to reduce flange climb derailment.
- Careful planning for a smooth transition to a new wheel profile.
- Optimizing gage distance on curves to improve vehicle curving and reduce wear.
- Searching for more effective techniques to deal with the leaf residue problem.
- Applying resilient rail/tie isolators to reduce noise.
- Investigating acceptable lubrication practices.
- Maintaining a high level of ride quality. Infrequent and somewhat transient hunting has been known to occur at certain places on the system, which seems to be driven by specific combinations of vehicle and track (including prevailing grade). As the vehicles age, this hunting situation may further deteriorate. If so, it is likely that a specific study of the cause/effect mechanisms will be required.

APPENDIX A-5
CTA

CTA operates the city's rapid transit system. Currently 225 mi of track are in service on seven lines, including about 25 mi of subway.

CTA uses four series of 48-foot long passenger cars similar in construction. These cars have a light axle load (19,200 pounds fully loaded) compared to other subway systems (around 26,000 pounds).

All but the oldest CTA cars use a large kingpin (~6-in. diameter and 14-in. length) allowing only rotational freedom about a vertical axis. All but the oldest cars have Wegmann style trucks, which equalize weight distribution by primary spring deflection and allowing the truck center plate to warp.

Routine car inspections are conducted at 6,000 mi or 90 days. Partial overhauls are conducted at one-quarter life, or about every 7 years. Complete overhauls are performed at half life, or about 12 to 14 years.

A5.1 WHEEL AND RAIL PROFILES

CTA uses 28-in. diameter wheels with the AAR narrow flange cylindrical tread profile with a flange angle of close to 68 degrees. This profile was adopted by CTA in the 1930s to eliminate vehicle hunting that occurred at 60 to 80 mph on high-speed, interurban cars.

Based on CTA's staff statement during the survey, this cylindrical profile has been performing well, likely because of two major factors:

- A high percentage of tangent track.
- A lighter axle load compared to other rapid transit.

CTA personnel recalled no mainline derailments in recent history. At the 54th Street yard, a few wheel-climb derailments have occurred in the past decade. These occurred at a 100-foot radius curve installed without a guardrail. The track worked well for 3 years, then an alignment of two factors caused a few climbs on newly re-profiled wheels:

- Acceptance of a new wheel re-profiling machine at the shop that might have increased the wheel surface roughness after re-profiling.
- Malfunction of the curve lubricator on this section of track.

The combination of these two factors could considerably reduce the L/V limit for flange climb on the tight curve that has a tendency to generate high lateral forces. Subsequently, a guardrail was installed and the lubricator has been regularly inspected.

The current rail profile at CTA is 115RE. However, short sections of older 90- and 100-lb/yd rail are still in use. Although starting with a crown radius, these rails are maintained to a flat head. Rail wear patterns identified in the field indicate that this wheel/rail combination has a rather wide contact band at wheel tread and rail head region.

A5.2 WHEEL LIFE AND WHEEL RE-PROFILING

CTA has not experienced flange wear problems using the AAR cylindrical tread contour. The wheel life is not actually tracked in routine maintenance. Thus, the current wheel life is unknown, but it was estimated as longer than 3 years and perhaps as much as 6 to 7 years.

CTA estimates that almost all wheel profile maintenance is done due to tread flat spots, mostly induced by operational causes such as braking, acceleration, and curving. Some flat spots may be caused by rail contamination, such as falling leaves. Subjective conditions for removing flats are mainly based on operator or public complaints. Since this is not a dimensional criterion, CTA personnel believe that a few flat spots approaching 2 in. in diameter can be found on the system. Occasionally, a wheel is re-profiled because it has exceeded the high flange limit. Wheel tread hollowing is rarely seen during wheel maintenance.

Wheel re-profiling is performed on lathe-style machines only at the Skokie shop and one Blue Line shop. CTA finds that this type of machine holds diameter variations on an axle much closer than a cutting head machine (within 0.005 to 0.010 in. from left to right wheel after re-profiling).

The current wheel diameter tolerances allowed after re-profiling are 3/64 in. within an axle (0.046 in.), 1 in. axle-to-axle in the same truck, and 1-in. truck-to-truck within the same car (i.e., using the same rule as above).

A5.3 RAIL LIFE AND RAIL GRINDING

Rails on tangent track and shallow curves have a relatively long life, perhaps 50 years. Further, rails last about 15 years even on the very tight curves at the corners of Chicago's famous Loop (89-foot radius). This is likely due to the low wheel loads, local rail lubrication, and low speeds (15 mph) allowed around the curves in the Loop.

CTA does not experience rail shelling because of light axle loading. Rail grinding is performed using a "rail smoother" that uses flat stones to grind a surface parallel to the top of the tie plates. The light grinding (using 8 to 10 grinder passes) is used on the whole system once a year to smooth corrugations and other imperfections.

New rails are installed with the original crown, and CTA smoothes the head about 1 year after installation. Typically after such smoothing, the rail needs no maintenance for another 4 years. Elevated track may require smoothing slightly more often.

A5.4 TRACK STANDARDS

Maximum speed on the CTA system today is 55 mph, with 15 mph limits on the tight (90-foot radius) Loop curves. The CTA track is generally designed to FRA Class 3 track geometry standards, but specifically applying the shorter 31-foot criteria. Also, CTA designs curves for a maximum of 4.5 in. of underbalance operation.

Track geometry is not measured regularly. A contractor was hired to measure the system in the early 1990s, but track geometry measurement has not been conducted since then. Visual inspections are performed twice a week by track walkers. No out-of-face, ultrasonic rail head inspections are performed, but ultrasonic bolthole/joint inspections are regularly scheduled.

All curves with a radius of less than 500 ft have guardrails, with a 1 7/8-in. flangeway clearance. The guardrail continues 10 ft before and after the curve. Designed track gage is the standard 56.5 in., except an additional 1/4-in. of width on curves tighter than a 125 ft radius. Maintenance is performed when the gage exceeds 1 in. in width.

A5.5 FIXATION METHODS

On elevated track, rails are fixed to full width wood ties. In the subway, CTA employs mostly wood half-ties in concrete. Some subway rails are held with coach screws, others with resilient fasteners. In and near the O'Hare Station, the track is directly fixed in concrete with resilient fasteners. CTA's surface tracks and a small amount of tunnel track are ballasted, with the use of spikes gradually being phased out in favor of clips.

To minimize dynamic car responses due to track stiffness variation at bridge approaches, CTA designed special 100-foot segments on either side of steel structures or bridges. Longer switch ties and closer-than-normal tie spacing were applied to create a stiffer support between the standard ballasted track and the elevated track.

CTA has noticed instances of the high rail lifting in locations that have greater than a 1 7/8-in. flangeway in a guarded curve. The greater flangeway width allowed higher lateral force against the high rail, and, consequently, the rail lifted from the tie plates. The rail was retained with standard cut spikes. The flangeway width was corrected to 1 7/8 in. and the problem was corrected.

A5.6 LUBRICATION AND WHEEL SLIDE

CTA uses both traditional wayside flange lubricators and a few trial wayside, top-of-rail friction modifier installations. Lubricators are installed on the curves with a radius of less than 500 ft. CTA track design personnel expressed interest in learning how different lubrication methods affect lateral forces.

The close spacing of trains on the system prevents the use of a hi-rail vehicle lubricator. No onboard rail lubrication is used or planned at this time. An earlier field trial of onboard solid stick flange lubricants (using the Skokie Swift Line) was unsatisfactory.

Regarding seasonal wheel slides, the Skokie Swift (Yellow Line) and the Brown Line near the end of the line on ballasted track have received noise complaints due to wheel flats. These flats are deemed to be related to leaf residue and resulting wheel slips. Removal of wheel flats is the active program to mitigate these complaints.

Other lines of the CTA system generally operate without leaf problems, likely due to two factors:

- Mostly underground, elevated, or freeway-median operating away from vegetation.
- Seven-day, 24-hour operations.

For these reasons, CTA does not have a rail-cleaning program.

A5.7 NOISE

Due to the extensive curve lubrication employed and current vehicle designs (lighter weights and relatively soft suspensions), CTA has reduced rail/wheel noise considerably in the past 25 years. CTA's solid wheels are now equipped with damper rings as shown in Figure A5-1, which has significantly reduced the free vibration of the plate.

Although the elevated structures remain rather noisy, further noise reductions are largely deemed impossible without reengineered support structures, since the steel girders are very effective noise amplifiers.

Figure A5-1. Steel damper on CTA wheels: The steel damper fits snugly into a groove on the field-side plate and reduces ringing considerably.

A5.8 MAJOR CONCERNS AND ACTIONS

CTA's wheel/rail profile maintenance focuses on keeping the wheels round and the rails smooth. Satisfactory vehicle designs have been proven over long periods, and CTA has no plans to change what works. Track installations and lubrication methods continue to evolve gradually, with a conservative policy toward trial installations.

Past problems have included the following:

- Vehicle hunting—improved by employing a cylindrical profile years ago.
- Curve squeal/screech—improved by slightly soft primary suspension trucks, light vehicles, rail lubrications, and lowered operating speeds.
- Slow-speed yard derailments—leading to greater attention on guardrail placement and lubricator maintenance.

CTA track designers would also like a computer program that allows input of design parameters (e.g., curve radius, operating speed, underbalance, lubrication presence, profile) and produces the expected lateral force ranges generated by the wheels.

APPENDIX B:

Investigation of Wheel Flange Climb Derailment Criteria for Transit Vehicles (Phase I Report)

INVESTIGATION OF WHEEL FLANGE CLIMB DERAILMENT CRITERIA FOR TRANSIT VEHICLES (PHASE I REPORT)

SUMMARY This research investigated wheelset flange climb derailment with the intent of developing limiting criteria for single-wheel L/V ratios and distance to climb for transit vehicles. The investigations used simulations of single wheelsets and representative transit vehicles. Based on the single wheelset simulation results, preliminary L/V ratio and climb-distance criteria for transit vehicle wheelsets are proposed. The proposed criteria are further validated through simulation of three types of transit vehicles. This research has been based on the methods previously used by the research team to develop flange climb derailment criteria for the North American freight railroads.

The following conclusions are drawn from single wheelset and vehicle simulations:

- New single wheel L/V distance criteria have been proposed for transit vehicles with specified wheel profiles:

Wheel 1 profile:

$$\text{L/V Distance (feet)} < \frac{5}{0.13 * \text{AOA} + 1}, \quad \text{if AOA} < 10 \text{ mrad}$$

$$\text{L/V Distance (feet)} = 2.2, \quad \text{if AOA} \geq 10 \text{ mrad}$$

Wheel 2 profile:

$$\text{L/V Distance (feet)} < \frac{4.1}{0.16 * \text{AOA} + 1}, \quad \text{if AOA} < 10 \text{ mrad}$$

$$\text{L/V Distance (feet)} = 1.6, \quad \text{if AOA} \geq 10 \text{ mrad}$$

Wheel 3 profile:

$$\text{L/V Distance (feet)} < \frac{4.2}{0.136 * \text{AOA} + 1}, \quad \text{if AOA} < 10 \text{ mrad}$$

$$\text{L/V Distance (feet)} = 1.8, \quad \text{if AOA} \geq 10 \text{ mrad}$$

Wheel 4/5 profile:

$$\text{L/V Distance (feet)} < \frac{28}{2 * \text{AOA} + 1.5}, \quad \text{if AOA} < 10 \text{ mrad}$$

$$\text{L/V Distance (feet)} = 1.3, \quad \text{if AOA} \geq 10 \text{ mrad}$$

Wheel 6 profile:

$$\text{L/V Distance (feet)} < \frac{49}{2 * \text{AOA} + 2.2}, \quad \text{if AOA} < 10 \text{ mrad}$$

$$\text{L/V Distance (feet)} = 2.2, \quad \text{if AOA} \geq 10 \text{ mrad}$$

where AOA is in mrad. In situations where AOA is not known and cannot be measured, the equivalent AOA (AOAe) calculated from curve curvature and truck geometry should be used in the above criteria.

- In situations where AOA is known and can be measured, more accurate new single wheel L/V ratio criteria based on AOA have also been proposed (see corresponding equation in Chapter 2 of this appendix).
- Simulation results for transit vehicles assembled with different types of wheel profiles confirm the validity of the proposed criteria.
- An incipient derailment occurs for most conditions when the climb distance exceeds the proposed criteria value.
- The proposed climb distance criteria are conservative for most conditions. Under many conditions, variations of AOA act to reduce the likelihood of flange climb.
- The single wheel L/V ratio required for flange climb derailment is determined by the wheel maximum flange angle, friction coefficient, and wheelset AOA.
- The L/V ratio required for flange climb converges to Nadal's value at higher AOA (above 10 mrad). For the lower wheelset AOA, the wheel L/V ratio necessary for flange climb becomes progressively higher than Nadal's value.
- The distance required for flange climb derailment is determined by the L/V ratio, wheel maximum flange angle, wheel flange length, and wheelset AOA.
- The flange climb distance converges to a limiting value at higher AOAs and higher L/V ratios. This limiting value is highly correlated with wheel flange length. The longer the flange length, the longer the climb distance. For the lower wheelset AOA, when the L/V ratio is high enough for the wheel to climb, the wheel-climb distance for derailment becomes progressively longer than the proposed flange-climb-distance limit. The wheel-climb distance at lower wheelset AOA is mainly determined by the maximum flange angle and L/V ratio.
- Besides the flange contact angle, flange length also plays an important role in preventing derailment. The climb distance can be increased through use of higher wheel maximum flange angles and longer flange length.
- The flanging wheel friction coefficient significantly affects the wheel L/V ratio required for flange climb. The lower the friction coefficient, the higher the single wheel L/V ratio required.
- For conventional solid wheelsets, a low nonflanging wheel friction coefficient has a tendency to cause flange climb at a lower flanging wheel L/V ratio, and flange climb occurs over a shorter distance for the same flanging wheel L/V ratio.
- The proposed L/V ratio and flange-climb-distance criteria are conservative because they are based on an assumption of a low nonflanging wheel friction coefficient.

- For independent rotating wheelsets, the effect of the nonflanging wheel friction coefficient is negligible because the longitudinal creep force vanishes.
- The proposed L/V ratio and flange-climb-distance criteria are less conservative for independent rotating wheels because independent rotating wheels do not generate significant longitudinal creep forces.
- For the range of track lateral stiffness normally present in actual track, the wheel-climb distance is not likely to be significantly affected by variations in the track lateral stiffness.
- The effect of inertial parameters on the wheel-climb distance is negligible at low speed.
- At high speed, the climb distance increases with increasing wheelset rotating inertia. However, the effect of inertial parameters is not significant at a low nonflanging wheel friction coefficient.
- Increasing vehicle speed increases the distance to climb.

Phase I of this project proposed specific L/V ratio and flange-climb-distance criteria for several specific wheel/rail profile combinations. Preliminary validation of these criteria was made using derailment simulations of several different passenger vehicles. To provide further validation of the criteria, the main task in Phase II of this project was to perform comparisons with results from full-scale transit vehicle tests. The conditions and limitations for the application of the criteria were also proposed.

Since the climb distance limit is highly correlated with the flange parameters (flange angle, length, and height), a general climb distance criterion that depends on both the AOA and flange parameters was further investigated in Phase II.

CHAPTER 1
INTRODUCTION

The research team conducted a full-scale wheel-climb derailment test with its TLV during 1994 and 1995 (*1*). The primary objective of the test was to reexamine the current flange climb criteria used in the Chapter XI track worthiness tests described in M-1001, *AAR Manual of Standards and Recommended Practices*, 1993.

In 1999, the research team conducted extensive mathematical modeling of a single wheelset flange using its dynamic modeling software (*2*). The objective of this work was to gain a detailed understanding of the mechanisms of flange climb. This research resulted in the proposal of a new single-wheel L/V ratio criterion and a new flange-climb-distance criterion for freight cars. Subsequently, some revisions were made to the proposed criteria (*3*).

Both of these projects were jointly funded by the FRA and the AAR.

The proposed L/V and distance-to-climb criteria were developed for freight cars with an AAR1B wheelset with a 75-degree flange angle. These were developed based on fitting L/V and distance-to-climb curves to numerous simulations of flange climb derailment. These were verified by comparison to the single wheel flange climb test results. Because the test and simulation results showed considerable sensitivity to axle AOA, the criteria were proposed in two forms. The first is for use when evaluating test results where the AOA is being measured, and the second, which is more conservative, is for use when the AOA is unknown or cannot be measured.

The following are the proposed criteria. Because measurement of AOA is usually quite difficult, the second forms are most likely to be used. The criteria are shown graphically in Figures B-1 and B-2.

(1) With capability to measure AOA during the test:

  (a) Wheel $\frac{L}{V} < 1.0$ {for AOA > 5 mrad}

  (b) Wheel $\frac{L}{V} < \frac{12}{\text{AOA (mrad)} + 7}$ {for AOA < 5 mrad}

(2) Without ability to measure AOA,

  Wheel $\frac{L}{V} < 1.0$

Correspondingly, the L/V distance criterion was proposed as:

(1) With onboard AOA measurement system,

  (a) L/V Distance (ft) $< \dfrac{16}{\text{AOA (mrad)} + 1.5}$

  {for AOA > −2 mrad}

  (b) L/V Distance (ft) $= \infty$ {for AOA < −2 mrad}

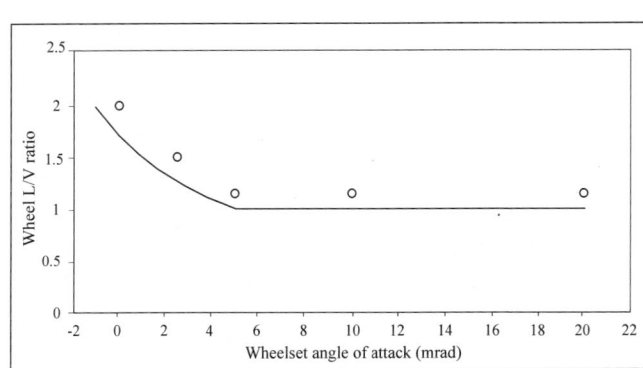

Figure B-1. Proposed single wheel L/V criterion with wheelset AOA measurement.

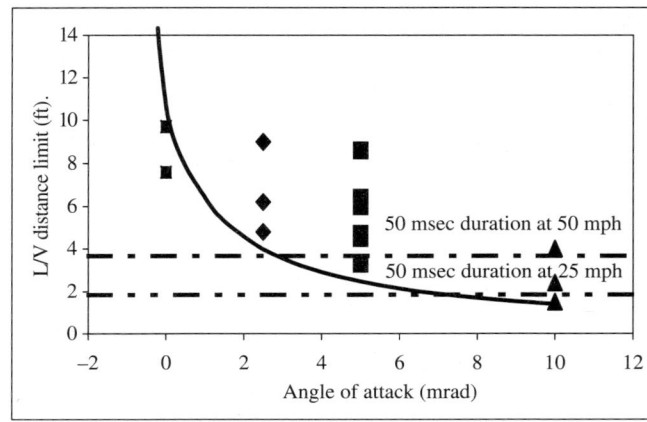

Figure B-2. Proposed L/V distance limit with wheelset AOA measurement. (Dots represent results; line represents the proposed distance limit.)

(2) Without onboard AOA measurement system, the L/V distance criterion is proposed relating to the track curvature:

$$\text{L/V Distance (ft)} < \frac{16}{\text{Curve (degree)} + 3.5}$$

The research to develop these criteria was based primarily on tests and simulations of wheel and rail profiles and loading conditions typical for the North American freight railroads. Analyses were also limited to 50 mph. The research team is conducting further research to finalize these proposed criteria for adoption by the AAR.

Currently, no consistent flange climb safety criteria exist for the North American transit industry. Wheel and rail profile standards and loading conditions vary widely for different transit systems and for different types of vehicles used in light rail and rapid transit services. ***Therefore, the proposed flange climb criteria developed by the research team for freight cars may not be directly applicable to any particular transit system***. The purpose of this project was to use similar analytical methods to develop flange climb derailment safety criteria, specifically for different types of transit systems and transit vehicles.

The research team undertook a program of developing wheel/rail profile optimization technology and flange climb criteria at the request of the NCHRP. This program included two phases, as listed Table B-1.

This report describes the methodology and results derived from the work performed in Task 2 of Phase I of this program. Wheel and rail profile data, and vehicle and track system data gathered as a part of Phase I, Task 1, were used to develop the inputs to the simulations of flange climb derailment.

1.1 BACKGROUND

Wheel-flange-climb derailments occur when the forward motion of the axle is combined with an excessive ratio of L/V wheel/rail contact forces. This usually occurs under conditions of reduced vertical force and increased lateral force that causes the wheel flange to roll onto the top of the rail head. The climb condition may be temporary, with wheel and rail returning to normal contact, or it may result in the wheel climbing fully over the rail. Researchers have been investigating the wheel flange climb derailment phenomena since the early 20th century. As a result of these studies, six flange climb criteria have been proposed. These criteria have been used by railroad engineers as guidelines for safety certification testing of railway vehicles. Briefly, they are the following:

- Nadal Single-Wheel L/V Limit Criterion
- Japanese National Railways (JNR) L/V Time Duration Criterion
- General Motors' Electromotive Division (EMD) L/V Time Duration Criterion
- Weinstock Axle-Sum L/V Limit Criterion
- FRA High-Speed Passenger Distance Limit (5 ft)
- AAR Chapter XI 50-millisecond (ms) Time Limit

The Nadal single-wheel L/V limit criterion, proposed by Nadal in 1908 for the French Railways, has been used throughout the railroad community. Nadal established the original formulation for limiting the L/V ratio in order to minimize the risk of derailment. He assumed that the wheel was initially in two-point contact with the flange point leading the tread. He concluded that the wheel material at the flange contact point was moving downwards relative to the rail material, due to the wheel rolling about the tread contact. Nadal further theorized that wheel climb occurs when the downward motion ceases with the friction saturated at the contact point. Based on his assumptions and a simple equilibrium of the forces between a wheel and rail at the single point of flange contact, Nadal proposed a limiting criterion as a ratio of L/V forces:

$$\frac{L}{V} = \frac{\tan(\delta) - \mu}{1 + \mu \tan(\delta)}$$

The expression for the L/V criterion is dependent on the flange angle δ and friction coefficient μ. Figure B-3 shows the solution of this expression for a range of values, appropriate to normal railroad operations. The AAR developed its

TABLE B-1 Wheel/rail profile optimization and flange climb criteria development tasks

Program: Development of Wheel/Rail Profile Optimization Technology and Flange Climb Criteria		
Phase I	Task 1	Survey the transit industry and define common problems and concerns related to wheel/rail profiles in transit operation
	Task 2	Propose preliminary flange climb derailment criteria for application to transit operation
Phase II	Task 1	Develop a general methodology of wheel/rail profile assessment applicable to transit system operation
	Task 2	Propose final flange climb derailment criteria validated by test data

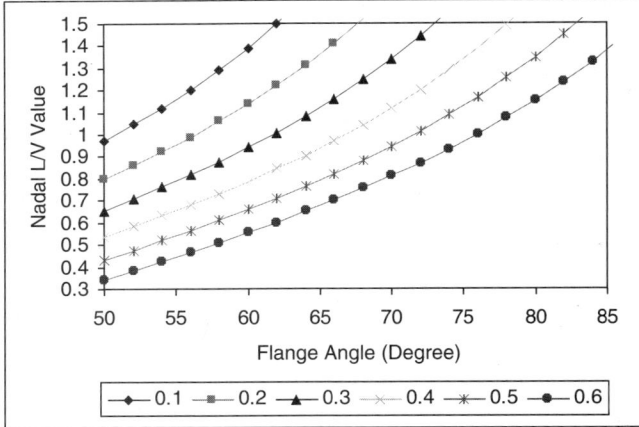

Figure B-3. Nadal criterion values.

Chapter XI single-wheel L/V ratio criterion based on Nadal's theory using a friction coefficient of 0.5.

Following a large number of laboratory experiments and observations of actual values of L/V ratios greater than the Nadal criterion at incipient derailment, researchers at JNR proposed a modification to Nadal's criterion (*4*). For time durations of less than 0.05 s, such as might be expected during flange impacts due to hunting, an increase was given to the value of the Nadal L/V criterion. However, small-scale tests conducted at Princeton University indicated that the JNR criterion was unable to predict incipient wheel-climb derailment under a number of test conditions.

A less conservative adaptation of the JNR criterion was used by General Motors EMD in its locomotive research (*5*).

More recently, Weinstock, of the United States Volpe National Transportation Systems Center, observed that this balance of forces does not depend on the flanging wheel alone (*6*). Therefore, he proposed a limit criterion that utilizes the sum of the absolute value of the L/V ratios seen by two wheels of an axle, known as the "Axle Sum L/V" ratio. He proposed that this sum be limited by the sum of the Nadal limit (for the flanging wheel) and the coefficient of friction (at the non-flanging wheel). Weinstock's criterion was argued to be not as overly conservative as Nadal's at small or negative AOA and less sensitive to variations in the coefficient of friction.

Based on the JNR and EMD research, and considerable experience in on-track testing of freight cars, a 0.05-s (50-ms) time duration was adopted by the AAR for the Chapter XI certification testing of new freight cars. This time duration has since been widely adopted by test engineers throughout North America for both freight and passenger vehicles.

A flange-climb-distance limit of 5 ft was adopted by the FRA for their Class 6 high speed track standards (*7*). This distance limit appears to have been based partly on the results of the joint AAR/FRA flange climb research conducted by the research team and also on experience gained during the testing of various commuter rail and long distance passenger cars.

A review of recent flange climb and wheel/rail interaction literature has been conducted as part of the work (shown in the Appendix B-1). Although several other teams are currently active in the field of flange climb research, no significant new flange climb criteria have been reported.

Therefore, it was concluded that the development of new criteria for the transit industry would be based on applying the research and analytical methods used in the research team's previous flange climb research. To develop wheel-climb derailment criteria for transit vehicles, some parameters—such as forward speed, inertial parameters of wheelsets, and wheel and rail profiles used in the transit industry—would need to be further investigated. The criteria also need to be further validated through simulations and tests of representative transit vehicles.

Previous research has also shown that flange climb is strongly influenced by wheelset AOA. Transit vehicles are likely to experience considerably different conditions of AOA than freight vehicles. Further, AOA is very difficult to measure. Thus, the proposed flange climb criteria are based on conservative expectations for AOA in different ranges of track curvature.

1.2 OBJECTIVE

The objectives of Phase I of this project were the following:

- To further investigate wheel/rail flange climb mechanisms for transit vehicles.
- To evaluate and propose wheelset flange climb derailment criteria for transit systems using simulations of single wheelsets.
- To validate the criteria through simulations of representative transit vehicles.

1.3 METHODOLOGY

1.3.1 Single Wheelset Flange Climb Derailment Simulations

The effects of different parameters on derailment were investigated through single-wheelset simulation. Based on these simulation results, the L/V ratio and climb distance criteria for six different kinds of transit wheelsets were proposed.

To minimize the number of variables and focus on wheel/rail interaction, a computer simulation model of a single wheelset was used. The wheel and rail profiles, inertia parameters, and vertical wheel loads were adopted from actual transit vehicle drawings and documents. Much of this data had been gathered as a part of the surveys being conducted for Phase I, Task 1 of this TCRP research project (*8*).

The same basic simulation methods used in the research team's previous flange climb studies were adopted here. To perform the flange climb derailment simulations, the wheelset AOA was set at a fixed value. A large yaw stiffness between the axle and ground ensured that the AOA remained approximately constant throughout the flange climb process. A vertical wheel load that corresponded to the particular vehicle axle load was applied to the wheelset to obtain the appropriate loading at the wheel/rail contact points.

The magnitude of the external lateral force and the wheelset AOA controls the flanging wheel L/V ratio. To make the wheel climb the rail and derail, an external lateral force was applied, acting towards the field side of the derailing wheel at the level of the rail head. Figure B-4 shows a typical lateral force history. During a constant speed movement, an initial lateral force was applied at either 50 percent or 80 percent of the expected L/V ratio for steady-state climb (based on Nadal's theory). This initial load level was held for 5 ft of travel to ensure equilibrium. The lateral force was then increased to the final desired L/V ratio (starting from A in Figure B-4). This high load was held until the end of the simulation. From this point, the wheel either climbed on top of the rail or it traveled a distance of 40 ft without flange climb; the latter was considered as no occurrence of derailment.

Flange climb results from each of the six different wheelsets were analyzed to develop and propose limiting flange climb L/V criteria and distance-to-climb criteria for the different types of transit systems.

1.3.2 Vehicle Derailment Simulations

As a preliminary validation of the proposed flange climb derailment criteria, three hypothetical passenger vehicles representing heavy rail and light rail transit vehicles were

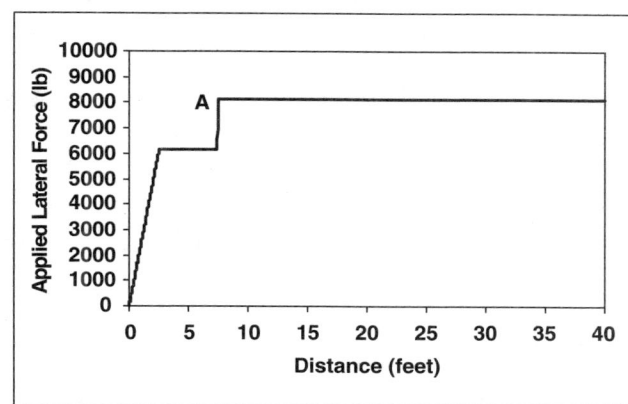

Figure B-4. Lateral force step input.

modeled. The vehicle models included typical passenger car components, such as air bag suspensions, primary rubber suspensions, and articulation joints.

To generate the large AOA, a large lateral force and vertical wheel unloading typical of actual flange climb conditions were used. The track input to the models used a measured track file, with variations in curvature, superelevation, gage, cross level, and alignment perturbations along the track.

The wheelset L/V ratio and climb distance for vehicles assembled with different wheelsets were obtained through vehicle simulations at different running speeds. The proposed flange climb derailment criteria were then evaluated by applying them to the vehicle simulation results.

CHAPTER 2
SINGLE WHEELSET FLANGE CLIMB DERAILMENT SIMULATIONS

The dynamic behavior of six different transit wheelsets were investigated through simulations of single wheelsets. The wheel profiles were taken from the transit system survey conducted as part of Phase I, Task 1 of this TCRP project. The basic parameters of these six wheelsets are listed in Table B-2.

Besides the wheel profiles, other parameters in the models—such as wheelset mass, inertia and axle loads—were adopted from drawings or corresponding documents to represent the real vehicle conditions in the particular transit systems.

For Wheels 1 through 5 (light rail and heavy rail) the simulations used a new AREMA 115 lb/yd rail section. For Wheel 6 (commuter rail) the AREMA 136 RE rail profile was used.

Flange climb results and the corresponding proposed limiting flange climb criteria are presented in the following sections for each of the six wheel profiles. A very detailed discussion is provided for Wheel 1. Since the same method was used for all profiles, only a synopsis of results is provided for the other five wheel profiles.

2.1 TRANSIT VEHICLE WHEELSET 1

Vehicle derailment usually occurs because of a combination of circumstances. Correspondingly, the indexes for the evaluation of derailment, the wheel L/V ratio, and climb distance are also affected by many factors. To evaluate the effects of these factors, case studies are presented for each of them in this section.

2.1.1 Definition of Flange Climb Distance

An important output parameter from the simulations is flange climb distance. The climb distance here is defined as the distance traveled from the final step in lateral force (point "A" in Figure B-4) to the point of derailment. For the purposes of

TABLE B-2 Wheel profile parameters

Parameter	Wheel 1	Wheel 2	Wheel 3	Wheel 4	Wheel 5	Wheel 6
Maximum Flange Angle (degree)	63.361	63.243	60.483	75.068	75.068	75.125
Nominal Wheel Diameter (in.)	28	27	27	26	26	36
Nadal Value	0.748	0.745	0.671	1.130	1.130	1.132
Flange Height (mm)	26.194	17.272	20.599	19.177	19.177	28.042
(in.)	1.031	0.680	0.811	0.755	0.755	1.104
Flange Length (mm)	19.149	11.853	17.232	10.038	10.038	15.687
(in.)	0.754	0.467	0.678	0.395	0.395	0.618
Source	WMATA	SEPTA-GRN	SEPTA-101	NJ-Solid	NJ-IRW	AAR-1B
Type of Service	Heavy rail	Light rail	Light rail	Light rail	Light rail, independent rotating wheels	Commuter cars

these studies, the point of derailment was determined by the contact angle on the flange tip decreasing to 26.6 degrees after passing the maximum contact angle of 63.3 degrees for Wheelset 1.

The 26.6-degree contact angle corresponds to the minimum contact angle for a friction coefficient of 0.5. Figure B-5 shows the wheel flange tip in contact with the rail at a 26.6-degree angle. Between the maximum contact angle (point Q) and the 26.6-degree flange tip angle (point O), the wheelset can slip back down the gage face of the rail due to its own vertical axle load if the external lateral force is suddenly reduced to zero. In this condition, the lateral creep force F (due to AOA) by itself is not large enough to cause the wheel to derail.

When the wheel climbs past the 26.6-degree contact angle (point O) on the flange tip, the wheelset cannot slip back down the gage face of the rail due to its own vertical axle load: the lateral creep force F generated by the wheelset AOA is large enough to resist the fall of the wheel and force the flange tip to climb on top of the rail.

As shown in Figure B-5, the flange length is defined as the sum of the maximum flange angle arc length QP and flange tip arc length PO.

2.1.2 Effect of Wheelset AOA

Figure B-6 shows the effect of AOA on wheel flange climb for Wheel 1 for a range of wheel L/V ratios. A friction coefficient of 0.5 was used on the flange and the tread of the derailing wheel.

Figure B-6 indicates that wheel climb will not occur for an L/V ratio less than the asymptotic value for each AOA. This asymptotic L/V value corresponds to the quasi-steady derailment value for this AOA. For L/V values higher than this, derailment occurs at progressively shorter distances. As AOA is decreased, the wheel quasi-steady derailment L/V value increases and the distance to climb also increases.

This result clearly indicates that the Nadal criterion is conservative for small AOAs, while for AOAs greater than 10 mrad flange climb occurs in distances less than 5 ft for L/V ratios that are slightly greater than the Nadal value.

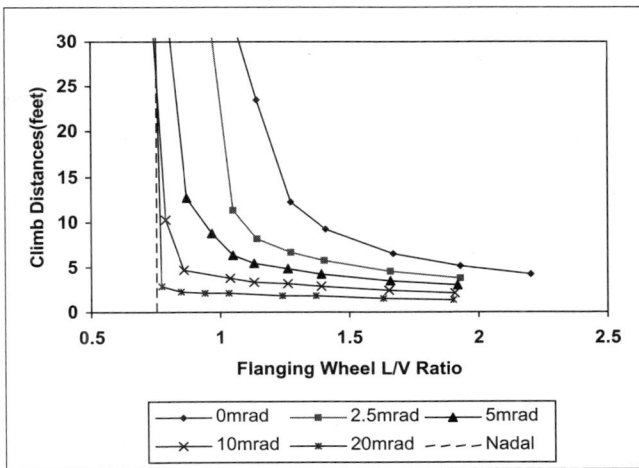

Figure B-6. Effect of wheelset AOA on distance to climb, $u = 0.5$ (Wheel 1).

2.1.3 Effect of Flanging Wheel Friction Coefficient

As indicated by Nadal's criterion (Figure B-3), the L/V ratio required for quasi-steady derailment is higher for a lower flanging wheel friction coefficient. Figure B-7 shows the effect on distance-to-climb of reducing the friction coefficient from 0.5 to 0.3. Compared with Figure B-6, the asymptotic L/V ratio for flanging wheel friction coefficient 0.3 in Figure B-7 is higher. However, as with the 0.5 coefficient of friction cases, for AOAs greater than 10 mrad flange climb still occurs in less than 5 ft for L/V ratios that are slightly greater than the Nadal value.

Figure B-8 compares the simulation results with Nadal's values for coefficients of friction of 0.1, 0.3, and 0.5 for a 5-mrad wheelset AOA. The dashed lines represent Nadal's values. The

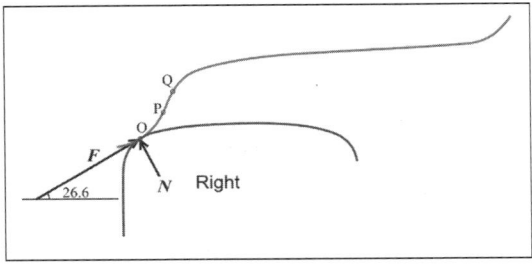

Figure B-5. Wheel/rail interaction and contact forces on flange tip.

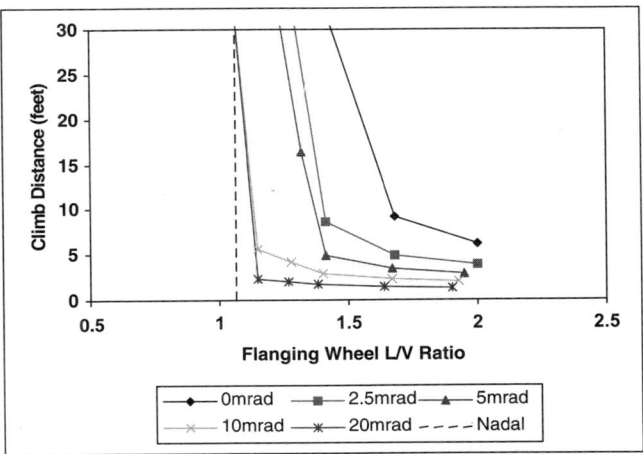

Figure B-7. Effect of wheelset AOA on distance to climb, $u = 0.3$ (Wheel 1).

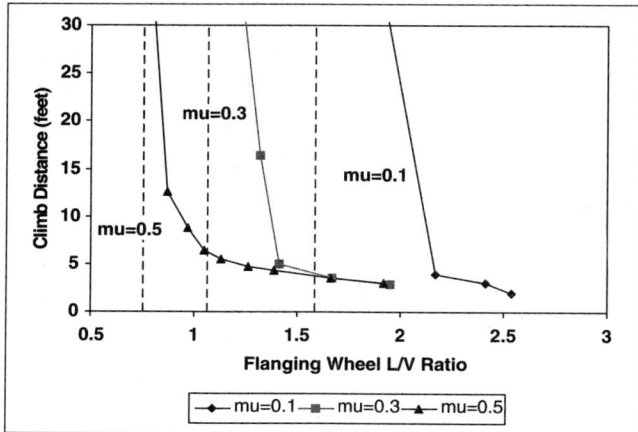

Figure B-8. Effect of coefficient of friction, 5 mrad AOA (Wheel 1).

asymptotic value increases with decreasing friction coefficient. A lower flange friction coefficient significantly increases the quasi-steady L/V ratio required for derailment but has almost no effect on the L/V distance limit if this L/V ratio is much exceeded.

2.1.4 Effect of Nonflanging Wheel Friction Coefficient

Figure B-9 shows the effect of the nonflanging wheel friction coefficient μ_{nf} with a flanging wheel friction coefficient of 0.5 and a 5-mrad wheelset AOA. At a very low μ_{nf}, the nonflanging wheel lateral and longitudinal creep forces were negligible; the initial high flanging wheel longitudinal creep force quickly decreased to the same small amplitude but in the reverse direction as the nonflanging longitudinal creep force.

When the nonflanging wheel friction coefficients are increased, the lateral creep forces on the nonflanging wheel side and the longitudinal creep forces on both sides become higher. As the longitudinal creep forces increase, the lateral creep force on the flanging wheel decreases with the saturation of resultant creep force. As a result, the quasi-steady wheel L/V ratio required for derailment increases, as shown in Figure B-9. However, if the L/V ratio is large enough to cause derailment (above 1.4 in Figure B-9), the climb distance is not affected by the nonflanging wheel friction coefficient.

This result indicates that a low nonflanging wheel friction coefficient has a tendency to cause flange climb at a lower flanging-wheel L/V ratio and climbs in a shorter distance than a wheelset with the same friction coefficient on both wheels. Low friction on the nonflanging wheel therefore represents the worst-case condition resulting in the shortest distances for flange climb. Thus, to produce conservative results, most of the single wheelset derailment simulations discussed in this report were performed with a very low nonflanging wheel friction coefficient (0.001).

2.1.5 Effect of Track Lateral Stiffness

Figure B-10 shows the effect of lateral track stiffness on the wheel flange climb at 5 mrad wheelset AOA. The difference of lateral track stiffness of 10^5 lb/in. and 10^6 lb/in. are negligible. As the lateral track stiffness decreases to 10^4 lb/in., the climb distance increases by 9 ft compared to the other two stiffness values. With stiffness of regular track normally in the range of 10^5 to 10^6 lb/in., the flange-climb distance is not likely to be significantly affected by the track lateral stiffness.

Note that the simulations do not allow the rail to roll. Therefore, the effect of reducing the track stiffness is to allow only increased lateral motion of the rails. In actual conditions of reduced lateral track stiffness it is common to have reduced

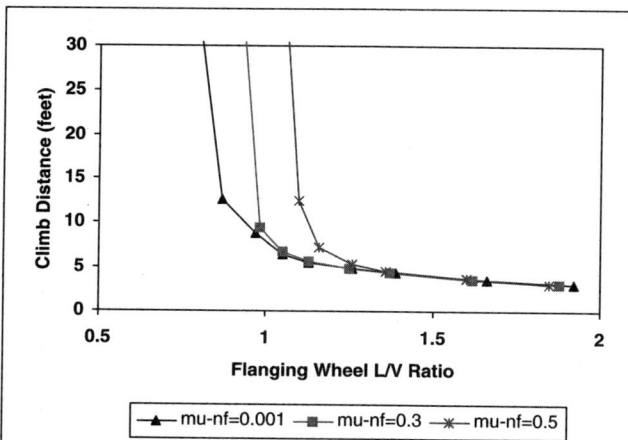

Figure B-9. Effect of nonflanging wheel friction coefficient, 5 mrad AOA (Wheel 1).

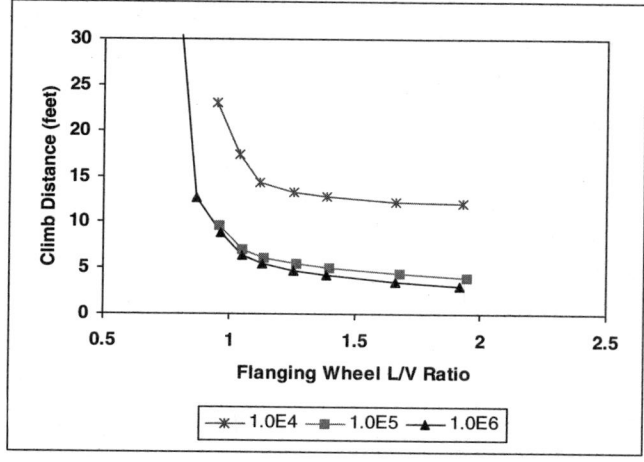

Figure B-10. Effect of track lateral stiffness, 5 mrad AOA (Wheel 1).

rail roll restraint as well. It is recommended that the effects of rail roll restraint and rail roll be studied at a future date.

2.1.6 Effect of Wheel L/V Base Level

As described in Section 1.3.1 (Figure B-4), all the simulations were performed with an initial lateral force applied to the wheelset to bring the wheelset towards flange contact and create a base L/V ratio. The base level represents the different L/V ratio that might be present due to steady state curving conditions or yaw misalignments of truck and/or axles in real vehicles.

Figure B-11 shows the effect of L/V ratio base level on the wheel flange climb at 5 mrad AOA. The climb distance decreases by a small amount as the L/V base level increases. The vertical dashed line in the plot is the quasi-steady L/V value for a 5-mrad AOA, which corresponds to the asymptotic line in Figure B-6. If the maximum L/V level is less than this value, flange climb derailment cannot occur for any L/V base level.

The climb distance is affected by the base L/V level, but the effect is much less significant than for a 75-degree-maximum-flange-angle wheelset (2). Compared to the low-flange-angle wheelset, the high-flange-angle wheelset requires greater effort to climb over the maximum flange angle and travels a farther distance at the same base L/V level. It is recommended that this difference be explored in greater detail in Phase II of this project. It is expected that the flange angle and the length of flange face that is maintained at the maximum contact angle will affect the climb distance.

2.1.7 Effect of Running Speed and Wheelset Inertial Parameters

Figure B-12 shows the effect of running speed on wheel flange climb at 5 mrad AOA. The climb distance increases with increases in running speed. As the stabilizing force

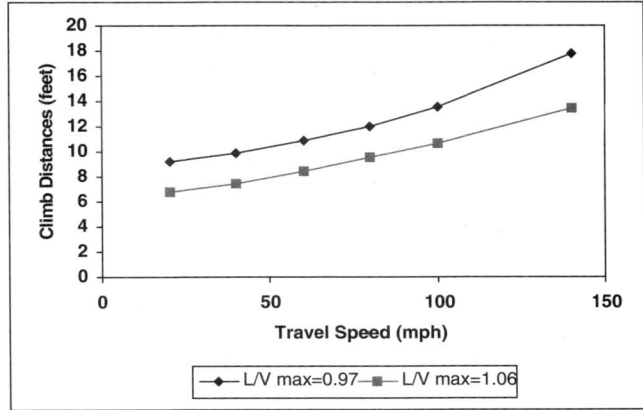

Figure B-12. Effect of speed on distance to climb for two different maximum wheel L/V ratios, 5 mrad AOA (Wheel 1).

for wheel derailment, a significant initial longitudinal creep force is generated at high speed and resists wheel climb.

The dynamic forces due to increased wheelset mass and rotating inertia become higher as the running speed increases. Figure B-13 shows the effect of inertial parameters on the climb distance for low speed (5 mph) and high speed (100 mph). The nominal wheelset rotating inertia was increased by two times, the wheelset mass and rolling and yawing inertia were also increased correspondingly. As seen in Figure B-13, at low speed, the effect of inertial parameters is negligible. At 100 mph, the climb distance of the double rotating inertia wheelset is increased by 0.5 to 1.0 ft at lower L/V ratios, but the effect of inertial parameters is negligible at high L/V ratios. The effect of inertia parameters is not significant at low nonflanging wheel friction coefficient.

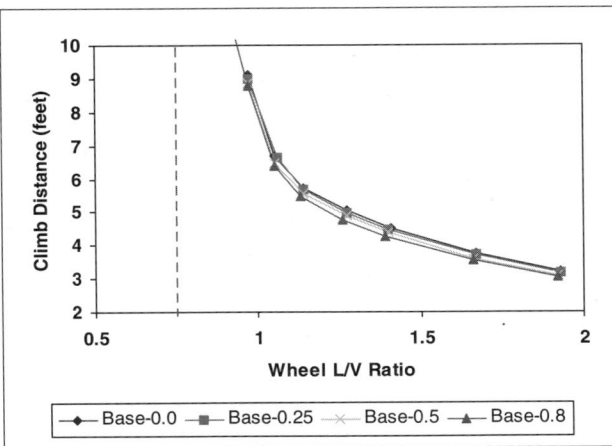

Figure B-11. Effect of L/V base level, 5 mrad AOA (Wheel 1).

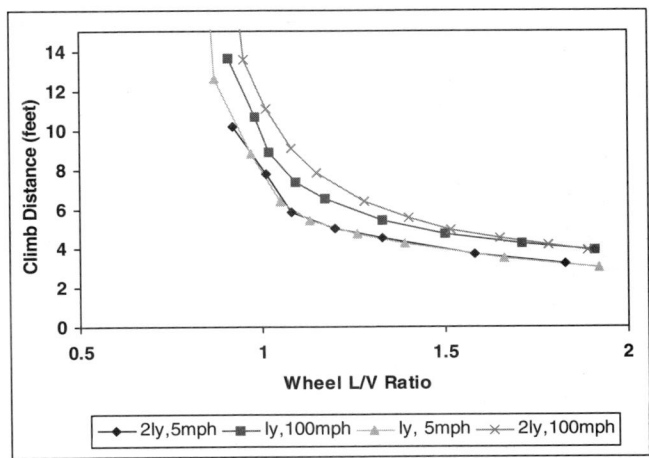

Figure B-13. Effect of rotating inertia at 5 mph and 100 mph, 5 mrad AOA (Wheel 1).

2.1.8 Wheel 1 Maximum Single Wheel L/V Ratio Criterion

Based on the above analysis, the AOA has the most significant effect on wheelset L/V ratio and climb distance. The following L/V ratio criteria for the Wheel 1 profile is proposed:

$$\frac{L}{V} < 0.74, \quad \text{if AOA} \geq 5 \text{ mrad} \quad (B-1)$$

$$\frac{L}{V} < \frac{11.3}{\text{AOA(mrad)} + 10.3}, \quad \text{if AOA} < 5 \text{ mrad} \quad (B-2)$$

$$\frac{L}{V} < 0.74, \quad \text{if AOA unknown} \quad (B-3)$$

Equations B-1 and B-3 are Nadal's limiting value for Wheel 1, which has a flange angle of 63 degrees. Equation B-2 was developed to account for the effects of increased flange climb L/V with small AOAs. This was done in a similar manner to previous TTCI research (2).

Figure B-14 shows the L/V ratio limit from the simulations compared to the proposed L/V ratio criterion for the Wheel 1 profile. Compared to the Nadal criterion, the new criterion is less conservative for AOA below 5 mrad. The proposed criterion, however, is more conservative than the simulations. This allows for the possibility of track and vehicle conditions that might cause localized increases in the AOA.

If a measurement of AOA is not available, a single value wheel L/V ratio criterion is proposed, as shown in Equation B-3.

2.1.9 Wheel 1 L/V Flange-Climb-Distance Criterion

The maximum wheel L/V ratio is constrained by the wheel L/V criterion proposed in Section 2.1.8. The maximum distance over which the L/V is permitted is given by the L/V distance criterion proposed below. The single wheelset simulation results, described above, show that the climb distance is strongly dependent on wheelset AOA. The following is the proposed L/V distance criterion for the Wheel 1 profile:

With onboard AOA measurement system,

$$\text{L/V Distance (feet)} < \frac{5}{0.13 * \text{AOA} + 1},$$

$$\text{if AOA} < 10 \text{ mrad} \quad (B-4)$$

$$\text{L/V Distance (feet)} = 2.2, \quad \text{if AOA} \geq 10 \text{ mrad} \quad (B-5)$$

Figure B-15 shows the simulation results of L/V climb distance and the proposed climb distance criterion for the Wheel 1 profile. All the simulation points are above the proposed criterion line. Therefore, the proposed flange-climb-distance criterion represents the worst case for all simulation cases and can be considered to be reasonably conservative.

When wheelset AOA is not available, an equivalent index AOA_e (in milliradians) of the leading axle of a two-axle truck can be obtained through a geometric analysis of truck geometry on a curve:

$$AOA_e = 0.007272clC \quad (B-6)$$

where
c = a constant for different truck types,
l = axle spacing distance in inches, and
C = the curve curvature, in degrees.

The relationship between the quasi-steady axle AOA on curve and the curve curvature was further investigated through a group of vehicle equilibrium position simulations. The vehicle model used is described below in Section 3.1.1.

Simulation results show the trailing axle of an H frame truck tends to align to a radial position while running through the curve. As a result, the leading axle AOA is increased correspondingly. When the curve curvature is larger than 4 degrees, the dynamic AOA/curvature ratio is

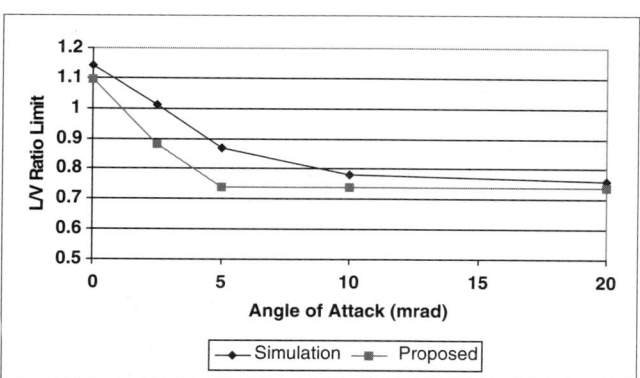

Figure B-14. Comparison of proposed wheel L/V ratio criterion with simulation (Wheel 1).

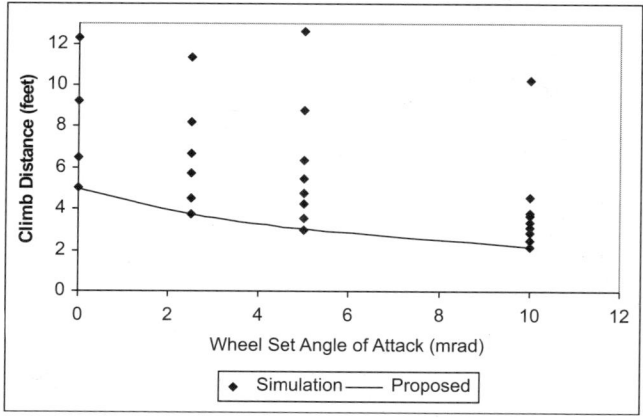

Figure B-15. Comparison of proposed L/V distance criterion with simulation (Wheel 1).

1–1.13, shown in Figure B-16, the constant c for the H frame truck is approximately 2 (1.8–2.1, as calculated by using Equation B-6 for the H frame truck with 75 in. axle spacing).

The simulation results in Figure B-16 also show the leading axle quasi-steady AOA value is smaller than the curve curvature at low curvature curve (below 5 degrees the ratio of AOA/curvature < 1), but larger than curvature at high curvature curve, the ratio of AOA/curvature > 1. Because the criterion is sensitive to low AOA, another constraint for Equation B-6 is added:

$$AOAe = C, \quad \text{if } C < 5 \text{ degree and } AOAe > C \quad (B\text{-}7)$$

For the situation without an onboard AOA measurement system, a criterion based on the track curvature is proposed:

$$\text{L/V Distance (feet)} < \frac{5}{0.13 * AOAe + 1} \quad (B\text{-}8)$$

AOAe (mrad) is the equivalent AOA calculated from curve curvature according to Equations B-6 and B-7.

2.2 TRANSIT VEHICLE WHEELSET 2

Figure B-17 shows the effect of wheelset AOA on wheel flange-climb distance for a range of Wheelset 2 flanging wheel L/V ratios. Coefficient of friction on the flanging wheel was 0.5. Results are generally similar to Wheel 1, with increased AOA requiring decreased distance to climb, although for each AOA the distances to climb are somewhat shorter.

2.2.1 Wheel 2 Maximum Single Wheel L/V Ratio Criterion

Based on the simulations, a proposed single wheel L/V criterion was developed for Wheel 2. Figure B-18 shows the NUCARS simulation L/V ratio limit and the proposed L/V ratio criterion for Wheel 2. The relationship of Wheel 2 L/V ratio versus AOA is quite similar to that of Wheel 1, because both wheels have the same 63-degree maximum flange angle. Therefore, the proposed L/V ratio criterion for the Wheel 2 is the same as that for Wheel 1 and is given by Equations B-1 to B-3.

2.2.2 Wheel 2 L/V Flange-Climb-Distance Criterion

Though the proposed L/V ratio criterion for Wheel 2 is the same as Wheel 1, the difference of its flange-climb distance is not negligible. The reasons are discussed in Section 2.7.

The following is the proposed L/V distance criterion for Wheel 2:

With onboard AOA measurement system,

$$\text{L/V Distance (feet)} < \frac{4.1}{0.16 * AOA + 1},$$
$$\text{if } AOA < 10 \text{ mrad} \quad (B\text{-}9)$$

$$\text{L/V Distance (feet)} = 1.6, \quad \text{if } AOA \geq 10 \text{ mrad} \quad (B\text{-}10)$$

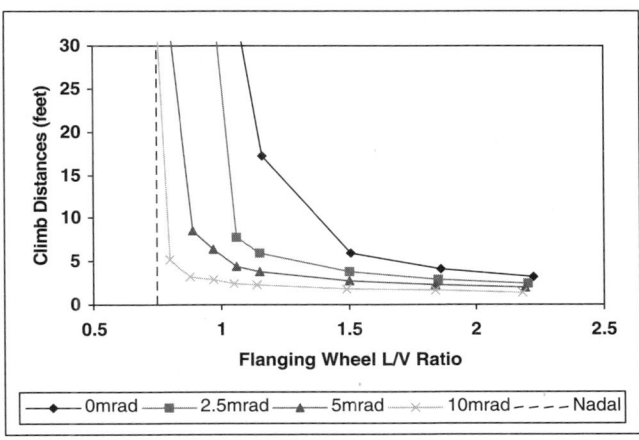

Figure B-17. Effect of wheelset AOA on distance to climb, u = 0.5 (Wheel 2).

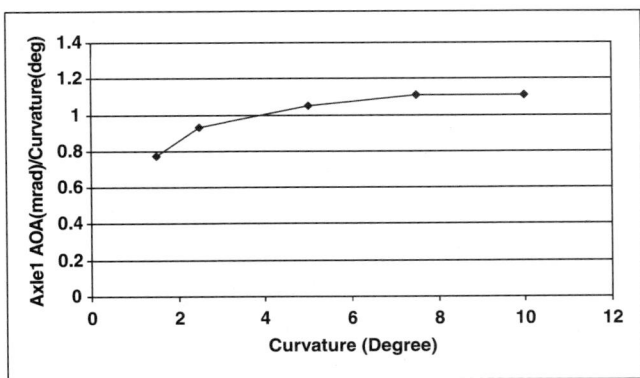

Figure B-16. Quasi-steady lead axle AOA as a function of curvature for an H-frame truck.

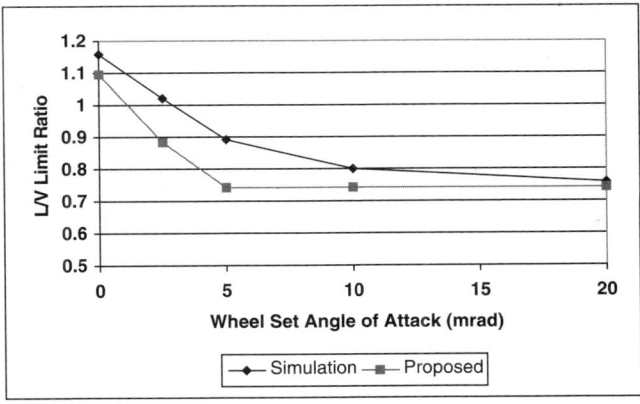

Figure B-18. Comparison of proposed wheel L/V ratio criterion with simulation (Wheel 2).

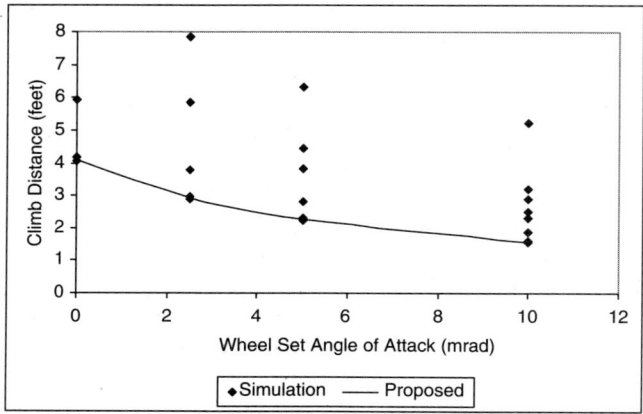

Figure B-19. Comparison of proposed L/V distance criterion with simulation (Wheel 2).

Without onboard AOA measurement system,

$$\text{L/V Distance (feet)} < \frac{4.1}{0.16 * AOAe + 1} \quad \text{(B-11)}$$

Figure B-19 shows the simulation results of L/V climb distance and the proposed climb distance criterion for Wheel 2.

2.3 TRANSIT VEHICLE WHEELSET 3

Figure B-20 shows the effect of wheelset AOA on flange-climb distance for a range of Wheelset 3 flanging wheel L/V ratios. The Nadal L/V flange climb limit is shown as a dashed line. Coefficient of friction on the flanging wheel was 0.5. Results are generally similar to Wheel 1, with increased AOA requiring decreased distance to climb. Distances to climb are slightly shorter and the Nadal limit is slightly lower due to the smaller flange angle of Wheel 3.

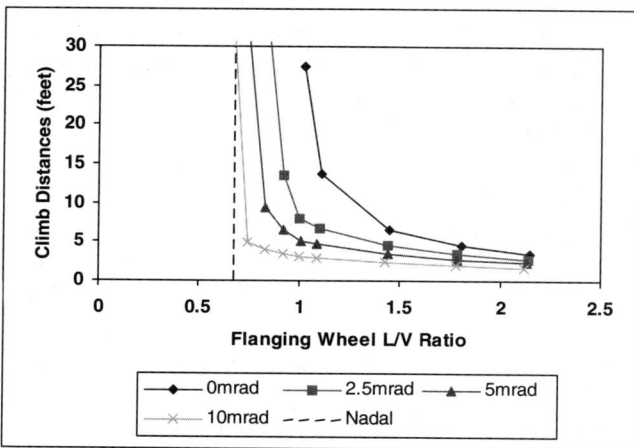

Figure B-20. Effect of wheelset AOA on distance to climb, u = 0.5 (Wheel 3).

2.3.1 Wheel 3 Maximum Single Wheel L/V Ratio Criterion

Based on the simulations, a proposed single wheel L/V criterion was developed for Wheel 3. This is different than for Wheel 1 and Wheel 2 due to the lower flange angle and corresponding lower Nadal limit. Figure B-21 shows the simulation L/V ratio limit and the proposed L/V ratio criterion for the Wheel 3. The following is the proposed L/V ratio criterion for Wheel 3:

$$\frac{L}{V} < 0.66, \quad \text{if AOA} \geq 5 \text{ mrad} \quad \text{(B-12)}$$

$$\frac{L}{V} < \frac{10.8}{AOA(mrad) + 11.38}, \quad \text{if AOA} < 5 \text{ mrad} \quad \text{(B-13)}$$

$$\frac{L}{V} < 0.66, \quad \text{if AOA unknown} \quad \text{(B-14)}$$

2.3.2 Wheel 3 L/V Flange-Climb-Distance Criterion

Figure B-22 compares simulation results of L/V climb distance and the proposed climb distance criterion for Wheel 3. The following is the proposed L/V distance criterion:

With onboard AOA measurement system,

$$\text{L/V Distance (feet)} < \frac{4.2}{0.136 * AOA + 1},$$
$$\text{if AOA} < 10 \text{ mrad} \quad \text{(B-15)}$$

$$\text{L/V Distance (feet)} = 1.8, \quad \text{if AOA} \geq 10 \text{ mrad} \quad \text{(B-16)}$$

Without onboard AOA measurement system,

$$\text{L/V Distance (feet)} < \frac{4.2}{0.136 * AOAe + 1} \quad \text{(B-17)}$$

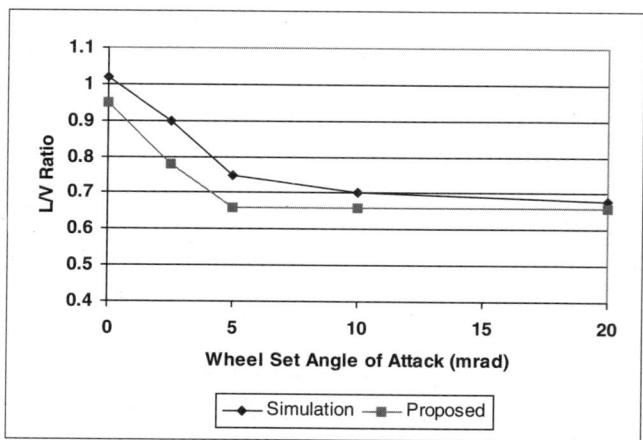

Figure B-21. Comparison of proposed wheel L/V ratio criterion with simulation (Wheel 3).

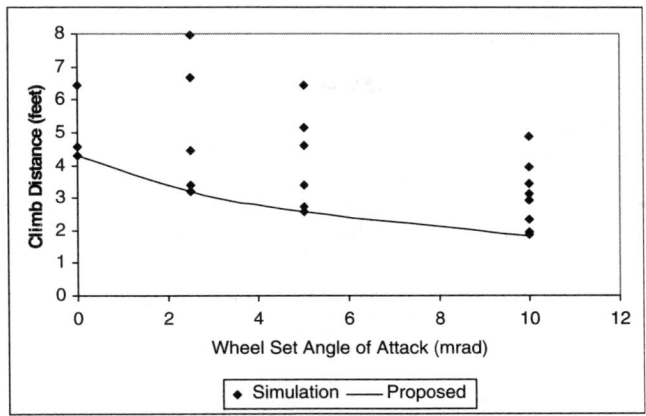

Figure B-22. Comparison of proposed L/V distance criterion with simulation (Wheel 3).

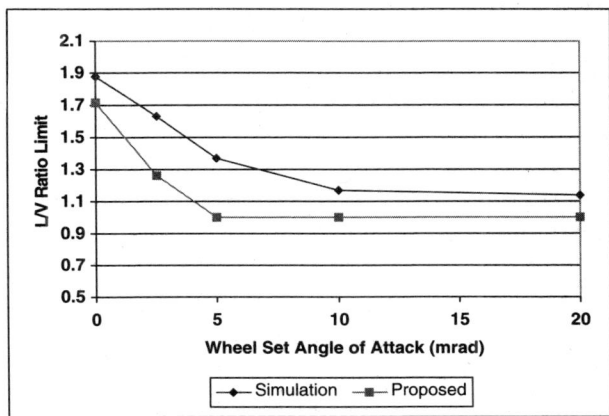

Figure B-24. Comparison of proposed wheel L/V ratio criterion with simulation (Wheel 4).

2.4 TRANSIT VEHICLE WHEELSET 4 (SOLID)

Figure B-23 shows the effect of wheelset AOA on flange-climb distance for a range of Wheelset 4 flanging wheel L/V ratios. The Nadal L/V flange climb limit is shown as a dashed line. Coefficient of friction on the flanging wheel was 0.5. Results are generally similar to Wheel 1, with increased AOA requiring decreased distance to climb. Although the Nadal limit is higher than for Wheel 1 due to a larger flange angle, the distances to climb are much shorter. This is due to a very short flange length, as discussed in Section 2.7.

2.4.1 Wheel 4 Maximum Single Wheel L/V Ratio Criterion

Based on the simulations, a proposed single wheel L/V criterion was developed for Wheel 4. This is different form Wheel 1 due to the larger flange angle and correspondingly higher Nadal limit. Due to the 75-degree flange angle, this criterion is the same as the L/V criterion proposed for freight vehicles (2). Figure B-24 shows the simulation L/V ratio limit and the proposed L/V ratio criterion for Wheel 4. The following is the proposed L/V ratio criterion for Wheel 4:

$$\frac{L}{V} < 1.0, \quad \text{if AOA} \geq 5 \text{ mrad} \quad \text{(B-18)}$$

$$\frac{L}{V} < \frac{12}{\text{AOA(mrad)} + 7}, \quad \text{if AOA} < 5 \text{ mrad} \quad \text{(B-19)}$$

$$\frac{L}{V} < 1.0, \quad \text{if AOA unknown} \quad \text{(B-20)}$$

2.4.2 Wheel 4 L/V Flange-Climb-Distance Criterion

Figure B-25 shows the simulation results of L/V climb distance and the proposed climb distance criterion for Wheel 4. It can be seen that some simulation points are below the proposed criterion line. However, bearing in mind that the L/V ratio of these points is much higher than the actual L/V ratio, which can be measured in practice, the proposed criterion is considered to be reasonable.

The following is the proposed L/V distance criterion:
With onboard AOA measurement system,

$$\text{L/V Distance (feet)} < \frac{28}{2 * \text{AOA} + 1.5},$$
$$\text{if AOA} < 10 \text{ mrad} \quad \text{(B-21)}$$

$$\text{L/V Distance (feet)} = 1.3, \quad \text{if AOA} \geq 10 \text{ mrad} \quad \text{(B-22)}$$

Without onboard AOA measurement system,

$$\text{L/V Distance (feet)} < \frac{28}{2 * \text{AOAe} + 1.5} \quad \text{(B-23)}$$

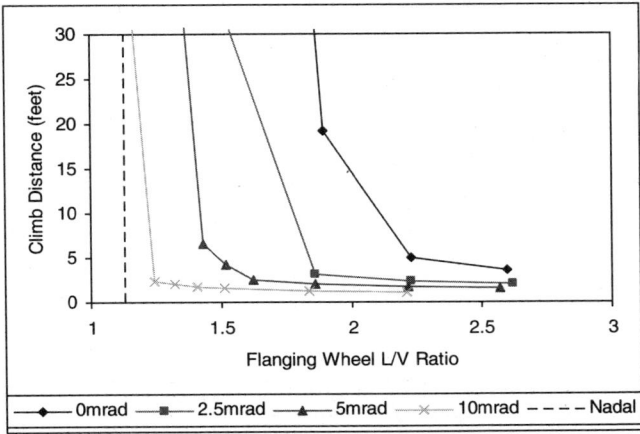

Figure B-23. Effect of wheelset AOA on distance to climb, u = 0.5 (Wheel 4).

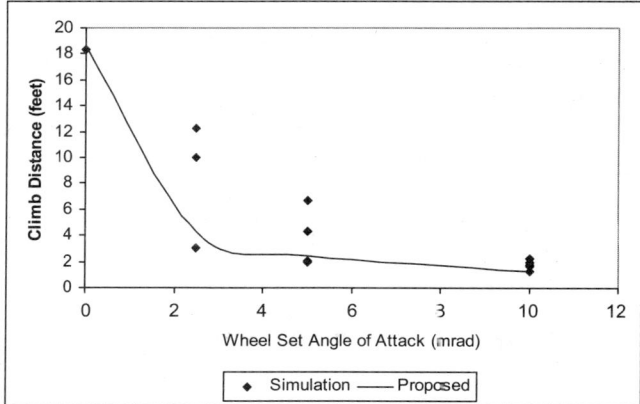

Figure B-25. Comparison of proposed L/V distance criterion with simulation (Wheel 4).

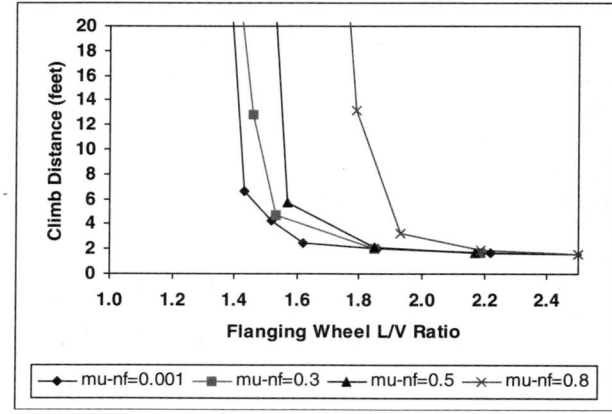

Figure B-26. Effect of nonflanging wheel friction coefficient, 5 mrad AOA (Wheel 4).

The proposed flange-climb-distance criterion for Wheel 4 is shorter than that proposed for the AAR-1B wheel (Section 2.6), which has the same flange angle. This is because the flange length on Wheel 4 is much shorter than for the AAR-1B wheel.

2.5 TRANSIT VEHICLE WHEELSET 5 (INDEPENDENT ROTATING WHEELS)

Wheel 4 and Wheel 5 have the same profile shape and flange angle. However, the left and right wheels of Wheelset 5 are allowed to rotate independently of each other, while Wheelset 4 has the wheels mounted on a solid axle. For Wheelset 5, this means the individual wheels are not constrained to rotate at the same speed, resulting in considerably different axle steering response.

2.5.1 Comparison of Solid and Independent Rotating Wheelsets

As discussed in Section 2.1.4, a low, nonflanging wheel friction coefficient represents the worst case for a conventional solid wheelset. Figure B-26 shows the effect of the nonflanging wheel friction coefficient μ_{nf} for Wheelset 4, with a flanging wheel friction coefficient of 0.5 and a 5-mrad wheelset AOA. With the increase of nonflanging wheel friction coefficients, the quasi-steady wheel L/V ratio required for derailment increases, and the climb distance also increases when the L/V ratio is lower than 2.2.

But the IRW wheelset situation is quite different because the spin constraint between the two wheels is eliminated. Figure B-27 shows the effect of the nonflanging wheel friction coefficient μ_{nf} for the independent rotating Wheelset 5, with a flanging wheel friction coefficient of 0.5 and a 5-mrad wheelset AOA. In contrast to the situation of the conventional solid wheelset, when $\mu_{nf} = 0.001, 0.3, 0.5,$ and 0.8, the longitudinal creep forces on both wheels vanish as expected. Therefore, longitudinal creep forces have no effect on lateral creep forces. For the independently rotating wheels at differ-

ent nonflanging wheel friction coefficient levels, the relationship between flanging L/V ratio and flange-climb distance converge to the same values.

Figure B-28 compares the conventional solid wheelset (Wheel 4) and independent rotating wheelset (Wheel 5) at different nonflanging wheel friction coefficients μ_{nf}. The difference is significant at large μ_{nf}. However, the results are virtually the same at small μ_{nf}, because for both cases there are almost no longitudinal creep forces present.

2.5.2 Wheel 5 Maximum Single Wheel L/V Ratio Criterion

The analyses in Section 2.5.1 show that the L/V ratios for independent rotating wheels and solid axles with low nonflanging wheel coefficient of friction are the same. Since Wheel 4 and Wheel 5 have the same wheel profile shape, the L/V criterion for Wheel 5 is the same as for Wheel 4 (Equations B-18 to B-20).

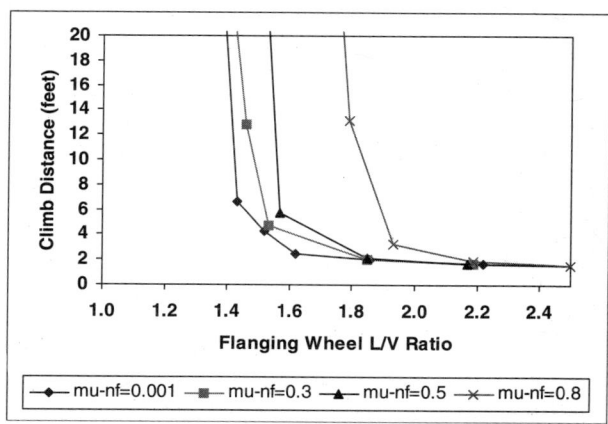

Figure B-27. Effect of nonflanging wheel friction coefficient, 5 mrad AOA (Wheel 5—independent rotating wheels).

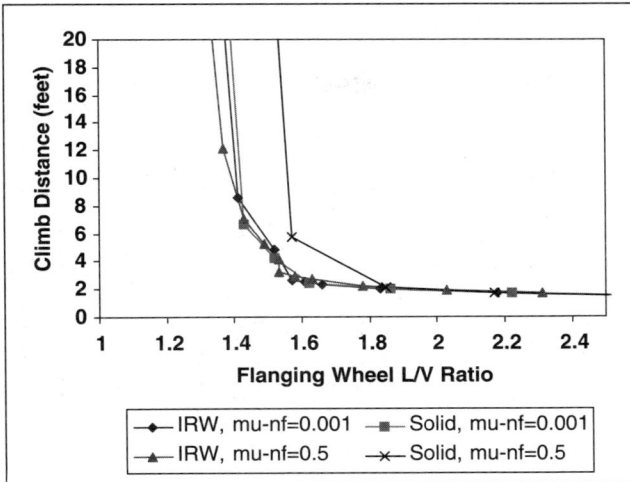

Figure B-28. Comparison of solid and IRW wheelset, 5 mrad AOA.

2.5.3 Wheel 5 L/V Flange-Climb-Distance Criterion L/V Distance Criterion

The analyses in Section 2.5.1 show that the distances to climb for independent rotating wheels and solid axles with low nonflanging wheel coefficient of friction are the same. Since Wheel 4 and Wheel 5 have the same wheel profile shape, the distance-to-climb criteria for Wheel 5 is the same as for Wheel 4 (Equations B-21 to B-23).

2.6 COMMUTER CAR WHEELSET 6

The commuter car wheelset uses the AAR-1B wheel profile, which has a 75-degree flange angle. Figure B-29 shows the effect of wheelset AOA on wheel flange-climb distance for a range of Wheelset 6 flanging wheel L/V ratios. The Nadal L/V flange climb limit is shown as a dashed line. Coefficient of friction on the flanging wheel was 0.5. Results are generally similar to Wheel 1, with increased AOA requiring decreased distance to climb. The Nadal limit is higher than for Wheel 1 due to a larger flange angle; and the distances to climb are somewhat longer, probably due to a longer flange length. The distances to climb for Wheel 6 are also longer than for Wheels 4 and 5, which also have the 75 degree flange angle. This is because of the longer flange length of the AAR-1B wheel profile.

2.6.1 Wheel 6 Maximum Single Wheel L/V Ratio Criterion

Based on the simulations, a proposed single wheel L/V criterion was developed for Wheel 6. This is different than for Wheel 1 due to the larger flange angle and corresponding higher Nadal limit. Because this is the AAR-1B wheel with a 75-degree flange angle, this criterion is the same as the L/V criterion proposed for freight vehicles, as well as for Wheels 4 and 5. Figure B-30 compares the simulation L/V ratio limit and the proposed L/V ratio criterion for Wheel 6. The proposed L/V ratio criterion for Wheel 6 is therefore given in Equations B-18 to B-20.

2.6.2 Wheel 6 L/V Flange-Climb-Distance Criterion

Figure B-31 shows the simulation results of L/V climb distance and the proposed climb distance criterion for Wheel 6. The following is the proposed L/V distance criterion:

With onboard AOA measurement system,

$$\text{L/V Distance (feet)} < \frac{49}{2 * \text{AOA} + 2.2}, \quad \text{(B-24)}$$
$$\text{if AOA} < 10 \text{ mrad}$$

$$\text{L/V Distance (feet)} = 2.2, \quad \text{if AOA} \geq 10 \text{ mrad} \quad \text{(B-25)}$$

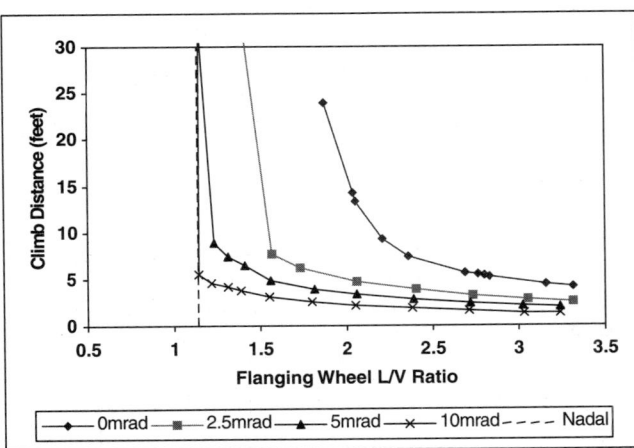

Figure B-29. Effect of wheelset AOA on distance to climb, u = 0.5 (Wheel 6).

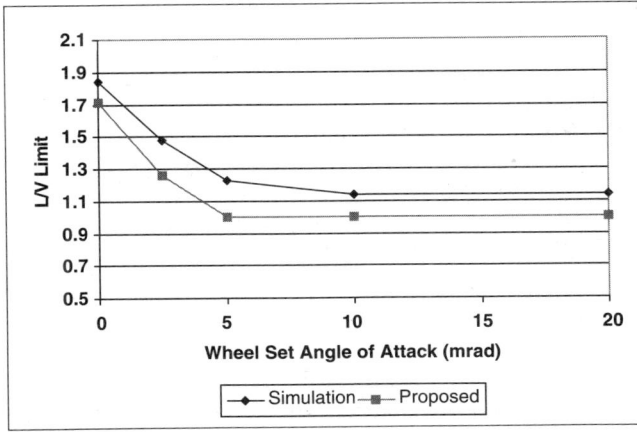

Figure B-30. Comparison of proposed wheel L/V ratio criterion with simulation (Wheel 6).

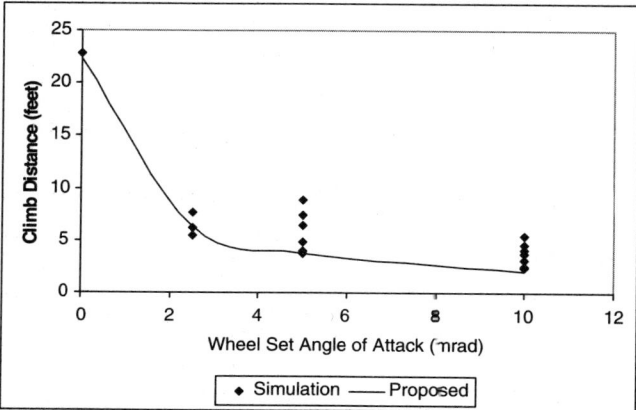

Figure B-31. Comparison of proposed L/V distance criterion with simulation (Wheel 6).

Without onboard AOA measurement system,

$$\text{L/V Distance (feet)} < \frac{49}{2 * AOAe + 2.2} \quad (B-26)$$

Although Wheel 6 is the same profile shape as used for freight cars, the proposed distance criterion is less conservative (longer distances) than the proposed distance criterion for freight cars. This is because the passenger vehicles normally have truck designs that control axle yaw angles better than standard freight cars and in some instances have softer primary suspensions that provide for better axle steering, resulting in lower AOA and longer flange climb distances.

2.7 COMPARISONS AND ANALYSIS

Figure B-32 compares all the above flange-climb-distance criteria and the proposed criterion for freight cars (3). As discussed in Section 2.1, the wheel L/V ratio decreases with the increase of wheelset AOA and converges to the Nadal value.

Figure B-32 shows that climb distances of all these different wheelsets decrease with increasing wheelset AOA and also converge to a corresponding asymptotic value. To understand the meaning of the asymptotic value, the wheel climbing process has to be examined in detail.

The wheelset climbing process is divided into two phases that are dependent on the wheel/rail contact positions. In the first phase, the flanging wheel contacts the rail at the maximum flange angle and begins to climb. The maximum flange angle is maintained for a certain length on the flange. For example, the length of Wheel 1 for maximum flange angle (63 degrees) is about 0.331 in. (8.4 mm). When the wheel climbs above the maximum flange angle, the wheel contacts the rail at the flange tip and begins the second climbing phase, with the contact angle reducing as the climb continues. The whole climbing process ends when the flange angle reaches 26.6 degrees. This is the point at which the wheelset can no longer fall back down the gage face of the rail by itself if the lateral flanging force is suddenly removed (corresponding to a friction coefficient of $\mu = 0.5$, as described in Section 2.1).

Figures B-33 and B-34 show the flanging wheel contact angle during climb and L/V ratio at 10 mrad AOA. The

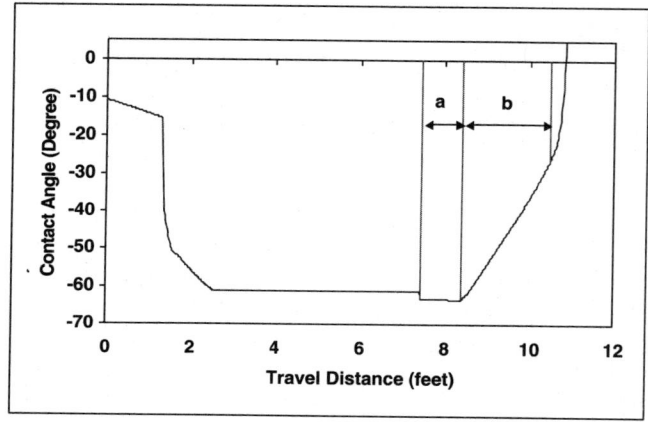

Figure B-33. Wheelset 1 contact angle, 10 mrad AOA.

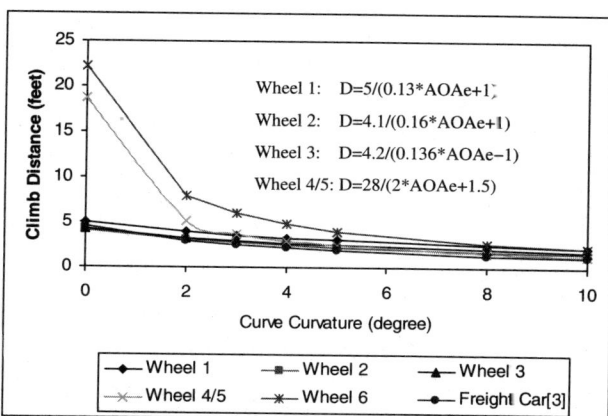

Figure B-32. L/V distance comparison for different wheelset profiles.

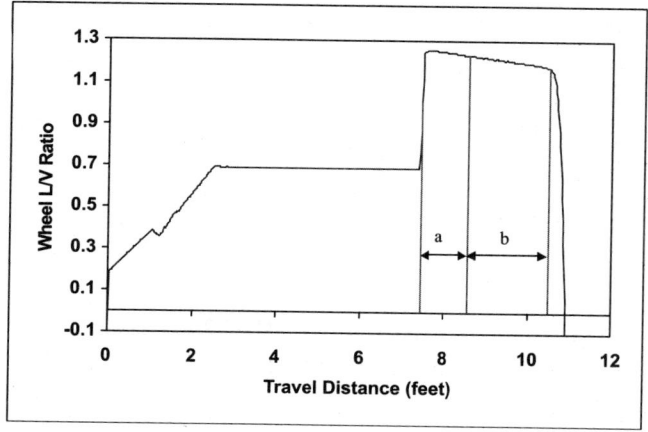

Figure B-34. Wheelset 1 L/V ratio, 10 mrad AOA.

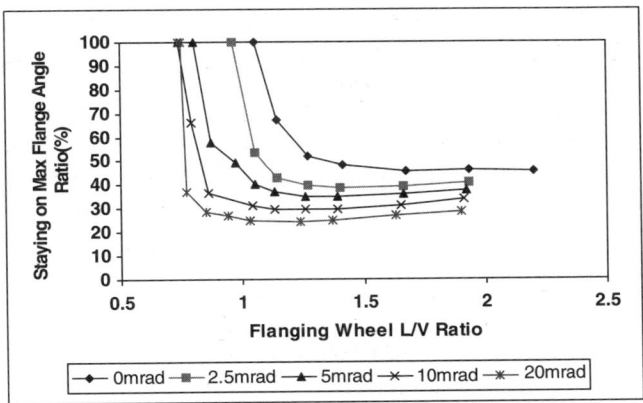

Figure B-35. Wheel 1 staying ratio with varying L/V ratio and AOA.

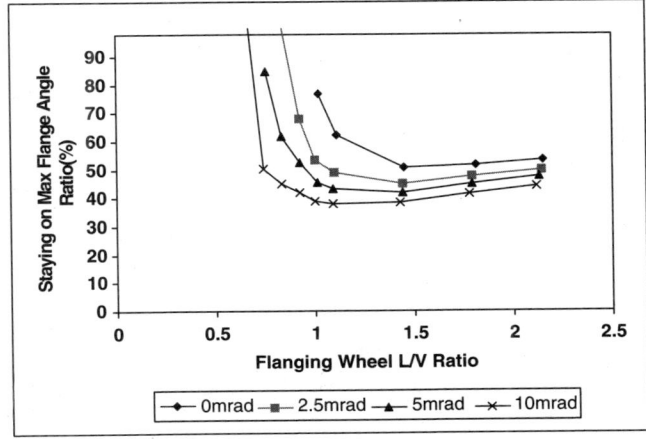

Figure B-37. Wheel 3 staying ratio with varying L/V ratio and AOA.

climbing distances are defined as "a" and "b" corresponding to the two climbing phases. In the first phase, where the contact stays on the maximum flange angle part of the flange (corresponding to arc QP in Figure B-5), a climbing ratio can be defined as a/(a+b). Figure B-35 shows that when the L/V ratio is greater than 1 and the AOA is greater than 10 mrad, the first phase climbing ratio is about 30 percent. This means that, for these conditions, 30 percent of the climb distance is on the maximum flange angle and 70 percent of the climb distance is on the flange tip.

As shown in Figures B-35 through B-39, for different wheel profiles, the whole climbing process is dominated by the phase of climbing on the flange tip (corresponding to arc PO in Figure B-5) for large wheelset AOA (above 5 mrad) and large L/V ratios (above 1.5 for 63 or 60 degrees maximum flange angle, above 2 for 75 degrees maximum flange angle). However, when the wheelset AOA is relatively small and the L/V ratio is lower, the whole climbing process is dominated by the first climbing phase and the wheelset mainly climbs on the maximum flange angle parts and then quickly derails.

Based on the above analysis, the wheel-climb distance is controlled by both the wheel/rail contact angle and the flange length, which includes the maximum flange angle length and the flange tip length. At a low AOA (<5 mrad) and a low L/V ratio, the wheel-climb distance is mainly determined by the maximum wheel flange angle—the higher the angle, the longer the climb distance. At a high AOA (>5 mrad) and a high L/V ratio, the wheel-climb distance converges to a limiting value. This limiting value is correlated with the wheel flange length, as demonstrated by the simulations. In very simple terms, the longer the flange length, the longer the climb distance limit.

A comparison of Nadal value and the proposed climb distance limits for the six different wheel profiles are listed in Table B-3.

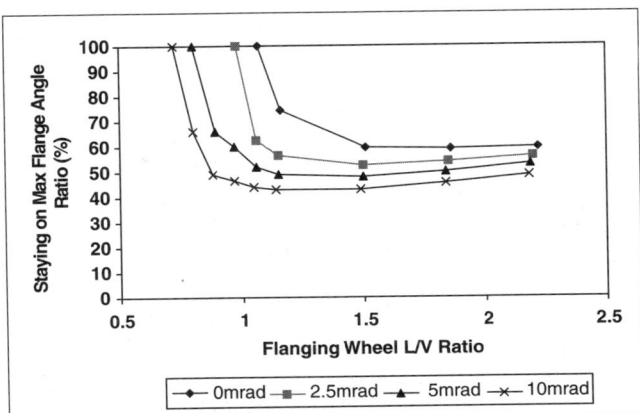

Figure B-36. Wheel 2 staying ratio with varying L/V ratio and AOA.

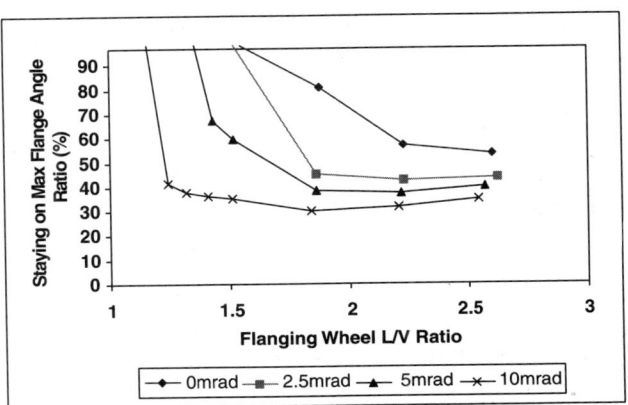

Figure B-38. Wheel 4 staying ratio with varying L/V ratio and AOA.

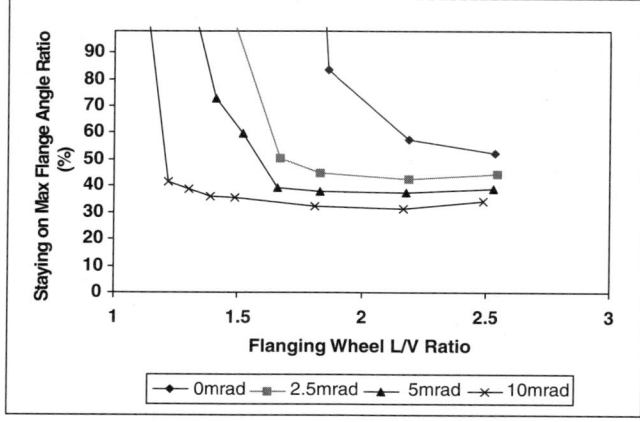

Figure B-39. Wheel 5 staying ratio with varying L/V ratio and AOA.

Figure B-40 shows the relationship between the flange-climb-distance limit and flange length. In general, the climb distance limit increases with increasing flange length. This conclusion means both the flange contact angle and the flange length play an important role in preventing derailment: increasing both the maximum flange angle and the flange length can increase the climb distance, thus improving derailment safety. The climb distance criterion, dependent on the AOA and flange parameters, will be further investigated in Phase II of this program.

2.8 CONCLUSIONS FOR SINGLE WHEELSET SIMULATIONS

Based on the single wheelset simulation results, wheel flange climb derailment criteria for transit vehicles have been proposed that are dependent on the particular wheel profile characteristics. The following conclusions can be drawn from the analyses performed:

- New single wheel L/V distance criteria have been proposed for transit vehicles with specified wheel profiles:

(1) Wheel 1 profile:

$$\text{L/V Distance (feet)} < \frac{5}{0.13 * \text{AOA} + 1}, \quad \text{if AOA} < 10 \text{ mrad}$$

$$\text{L/V Distance (feet)} = 2.2, \quad \text{if AOA} \geq 10 \text{ mrad}$$

(2) Wheel 2 profile:

$$\text{L/V Distance (feet)} < \frac{4.1}{0.16 * \text{AOA} + 1}, \quad \text{if AOA} < 10 \text{ mrad}$$

$$\text{L/V Distance (feet)} = 1.6, \quad \text{if AOA} \geq 10 \text{ mrad}$$

(3) Wheel 3 profile:

$$\text{L/V Distance (feet)} < \frac{4.2}{0.136 * \text{AOA} + 1}, \quad \text{if AOA} < 10 \text{ mrad}$$

$$\text{L/V Distance (feet)} = 1.8, \quad \text{if AOA} \geq 10 \text{ mrad}$$

(4) Wheel 4/5 profile:

$$\text{L/V Distance (feet)} < \frac{28}{2 * \text{AOA} + 1.5}, \quad \text{if AOA} < 10 \text{ mrad}$$

$$\text{L/V Distance (feet)} = 1.3, \quad \text{if AOA} \geq 10 \text{ mrad}$$

(5) Wheel 6 profile:

$$\text{L/V Distance (feet)} < \frac{49}{2 * \text{AOA} + 2.2}, \quad \text{if AOA} < 10 \text{ mrad}$$

$$\text{L/V Distance (feet)} = 2.2, \quad \text{if AOA} \geq 10 \text{ mrad}$$

TABLE B-3 Nadal values and climb distance limits for different wheelset profiles

Items	Wheel 1	Wheel 2	Wheel 3	Wheel 4/5	Wheel 6
Maximum Flange Angle (Degree)	63.361	63.243	60.483	75.068	75.125
Nadal Value	0.748	0.745	0.671	1.130	1.132
Flange Length (mm)	19.149	11.853	17.232	10.038	15.687
(in.)	0.754	0.467	0.678	0.395	0.618
Climb Distance in feet (at 10 mrad AOA)	2.174	1.577	1.780	1.302	2.207

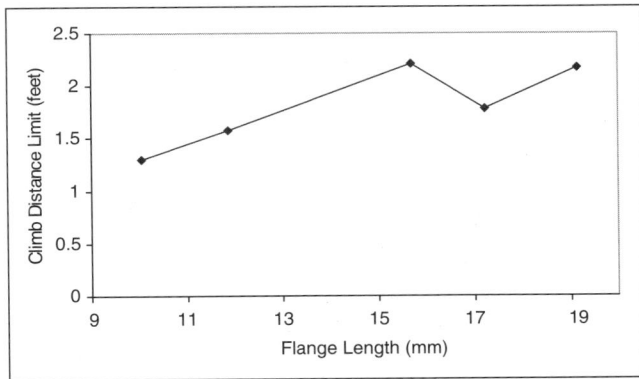

Figure B-40. Relation between the climb distance limit and the flange length.

where AOA is in mrad. In situations where AOA is not known and cannot be measured, the equivalent AOA (AOAe) calculated from curve curvature and truck geometry should be used in the above criteria.

- If the AOA is known and can be measured, more accurate new single wheel L/V ratio criteria based on AOA have also been proposed (see corresponding equation in Chapter 2).
- The single wheel L/V ratio required for flange climb derailment is determined by the wheel maximum flange angle, friction coefficient, and wheelset AOA.
- The L/V ratio required for flange climb converges to Nadal's value at higher AOAs (above 10 mrad). For lower wheelset AOAs, the wheel L/V ratio necessary for flange climb becomes progressively higher than Nadal's value.
- The distance required for flange climb derailment is determined by the L/V ratio, wheel maximum flange angle, wheel flange length, and wheelset AOA.
- The flange-climb distance converges to a limiting value at higher AOAs and higher L/V ratios. This limiting value is highly correlated with wheel flange length. The longer the flange length, the longer the climb distance.
- For lower wheelset AOAs, when the L/V ratio is high enough for the wheel to climb, the wheel-climb distance for derailment becomes progressively longer than the proposed flange-climb-distance limit. The wheel-climb distance at lower wheelset AOAs is mainly determined by the maximum flange angle and L/V ratio.
- Besides the flange contact angle, flange length also plays an important role in preventing derailment. The climb distance can be increased through the use of higher wheel maximum flange angles and longer flange length.
- The flanging wheel friction coefficient significantly affects the wheel L/V ratio required for flange climb: the lower the friction coefficient, the higher the single wheel L/V ratio required to climb the rail.
- For conventional solid wheelsets, a low nonflanging wheel friction coefficient has a tendency to cause flange climb at a lower flanging wheel L/V ratio. Flange climb occurs over a shorter distance for the same flanging wheel L/V ratio.
- The proposed L/V ratio and flange-climb-distance criterion are conservative because they are based on an assumption of a low nonflanging wheel friction coefficient.
- For independent rotating wheelsets, the effect of the nonflanging wheel friction coefficient is negligible because the longitudinal creep force vanishes.
- The proposed L/V ratio and flange-climb-distance criterion are less conservative for independent rotating wheels because independent rotating wheels do not generate significant longitudinal creep forces.
- For the range of track lateral stiffness normally present in actual track, the wheel-climb distance is not likely to be significantly affected by variations in the track lateral stiffness.
- The effect of inertial parameters on the wheel-climb distance is negligible at low speeds.
- At high speeds, the climb distance increases with increasing wheelset rotating inertia. However, the effect of inertial parameters is not significant at a low nonflanging wheel friction coefficient.

CHAPTER 3

TRANSIT VEHICLES FLANGE CLIMB DERAILMENT SIMULATIONS

An AOA measurement is not usually available in practice. Therefore, the proposed criteria based on curvature will normally be used. These are validated in this section of this report. Three types of hypothetical passenger cars representing heavy rail and light rail transit vehicles have been modeled. The proposed criteria were applied to the simulation results to evaluate the validity of the proposed criteria.

The vehicle models include typical passenger car connections, such as air bag suspensions and articulated units. The wheel/rail connection parameters in the vehicle models correspond to the values adopted in the single wheelset model in Chapter 2. Other suspension parameters were estimated according to the research team's previous experience in simulation of passenger vehicles.

3.1 HEAVY RAIL VEHICLE

3.1.1 The Vehicle Model

The vehicle modeled is a typical heavy rail transit system vehicle. It has H-frame trucks, chevron primary suspension and secondary air suspension. The principal dimensions of the car are as follows: (1) car length over couplers 67 ft 10 in., (2) rigid wheel base 6 ft 10 in., (3) wheel diameter 28 in., and (4) truck centers 47 ft 6 in. A loaded car weight of 108,664 pounds was used to calculate the required car body mass and mass moment of inertias for the vehicle model. Overall, a total of 12 bodies and 60 connections were used to assemble the simulation model.

3.1.2 Track Geometry Input Data

The track input to the simulations comprises track geometry data and track curve data. The track geometry data is used to specify perturbed track input to the model and consists of lateral and vertical perturbation amplitudes of each rail at specified positions along the track. The track curve data is used for specifying the superelevation and curvature of the track.

For these simulations, measurements of actual track with a large alignment and cross level perturbation in a curve were adopted because they were expected to generate conditions that could lead to wheel flange climb derailment. The input data included the curvature, superelevation, gage, cross level, profile, and alignment perturbations along the track. Figures B-41 through 46 show the curvature, superelevation, and perturbations. A large drop in left rail vertical direction, alignment, and cross level perturbation can be found at the 558-ft distance.

The track geometry coordinate system follows right-hand rules. The longitudinal axis is parallel to the track centerline, with positive displacements in the direction of travel. The lateral axis is perpendicular to the track centerline and is positive pointing to the left, when viewed in the direction of motion. The vertical axis is positive pointing upward to

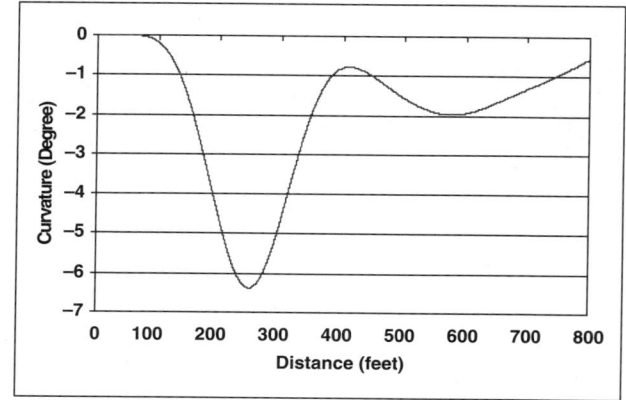

Figure B-41. Track curvature.

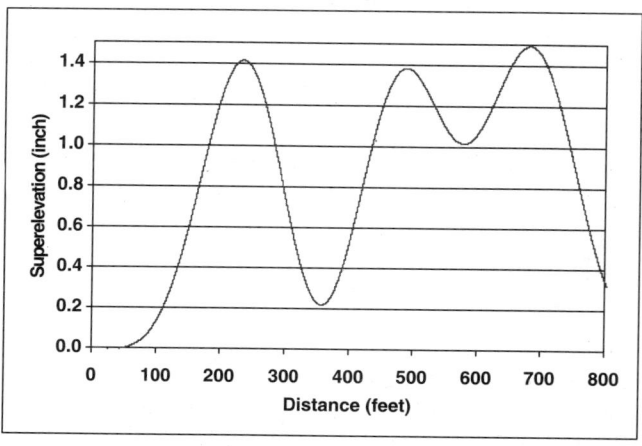

Figure B-42. Track superelevation.

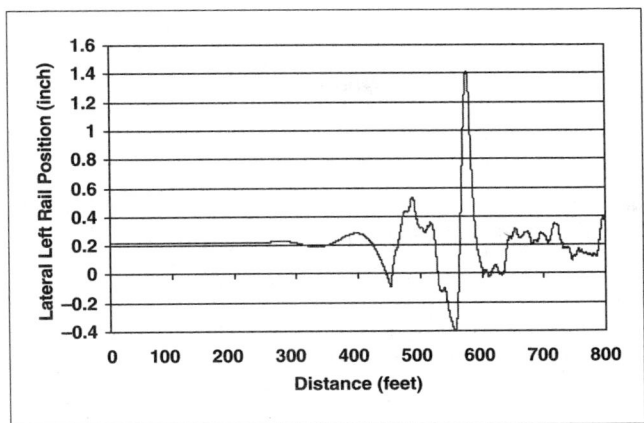

Figure B-43. Left rail lateral position.

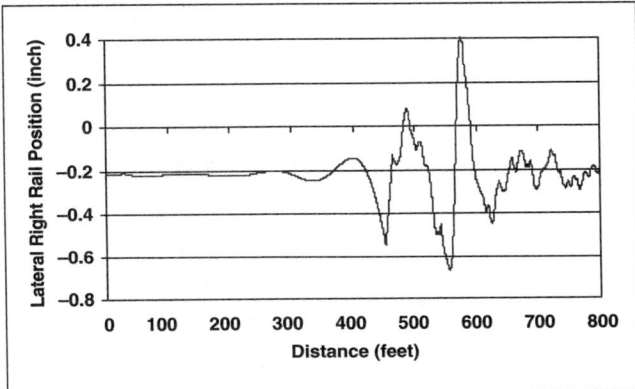

Figure B-44. Right rail lateral position.

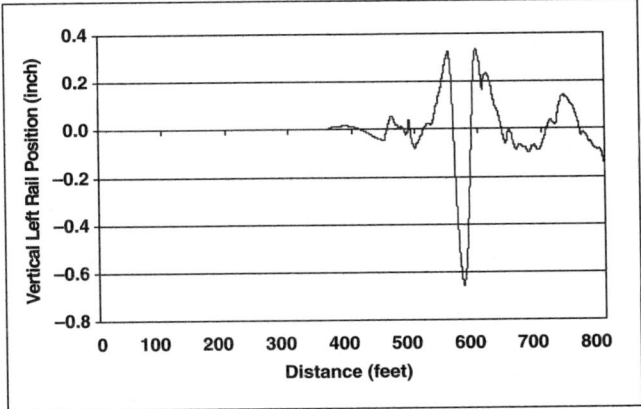

Figure B-45. Left rail vertical position.

complete the coordinate system. Negative curvature indicates a right-hand curve; positive superelevation is for a superelevated right-hand curve. Correspondingly, the left wheel climbs on the left rail while the vehicle travels through the right-hand curve.

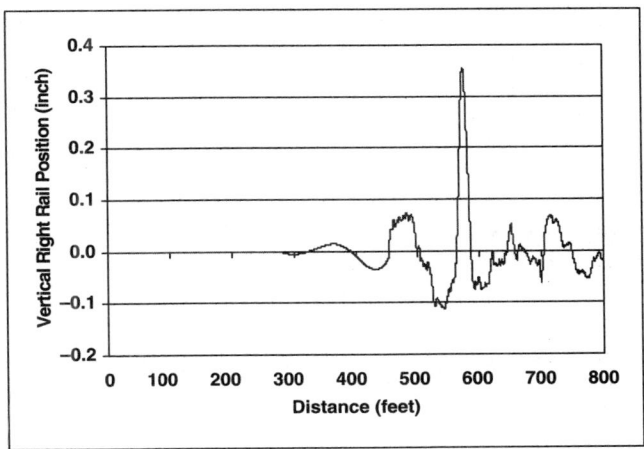

Figure B-46. Right rail vertical position.

3.1.3 Heavy Rail Vehicle Simulation Results

Simulations for the heavy rail vehicle were made for two different wheel profiles to validate the corresponding proposed L/V ratio and distance-to-climb criteria.

The first wheel/rail profile combination used was Wheel 1 on standard AREMA 115-10 lb/yd rail. These are the same as used for the Wheel 1 single wheelset simulations discussed in Section 2.1. The rail/wheel friction coefficient is 0.5.

The second wheel/rail profile combination used was Wheel 2 on standard AREMA 115-10 lb/yd rail. These are the same as used for the Wheel 2 single wheelset simulations discussed in Section 2.2. The rail/wheel friction coefficient is 0.5.

Both wheel profiles have a 63-degree flange angle, but Wheel 2 has a much shorter flange length and a correspondingly shorter distance-to-climb criterion.

The simulations were conducted for a range of speeds to generate a range of flange climbing conditions:

- Contact with maximum flange angle but not flange climbing.
- Flange beginning to climb up the rail but not derailing (incipient derailment).
- Flange climbing that terminated in derailment.

This range of conditions represents what happens to actual vehicles when they encounter severe track perturbations. The proposed criteria were evaluated by comparing them to the results for these different flange climb conditions.

3.1.3.1 Heavy Rail Vehicle Assembled with Wheelset 1

Simulation results show that the first axle begins to climb at a track location of 555.5 ft (distance referred to the first axle). This is the location of the large lateral and vertical track geometry perturbations. Distance histories of wheel L/V ratio along the track at speeds of 40, 50, and 60 mph are given in Figure B-47. As the speed increases, the high L/V ratio is sustained for a longer distance. This results in the wheel being at the maximum flange angle for a longer distance, as shown in Figure B-48, corresponding to a longer flange climb distance.

For Wheelset 1, the Nadal value is 0.748. According to Equation B-3, the proposed L/V ratio limiting value is 0.74.

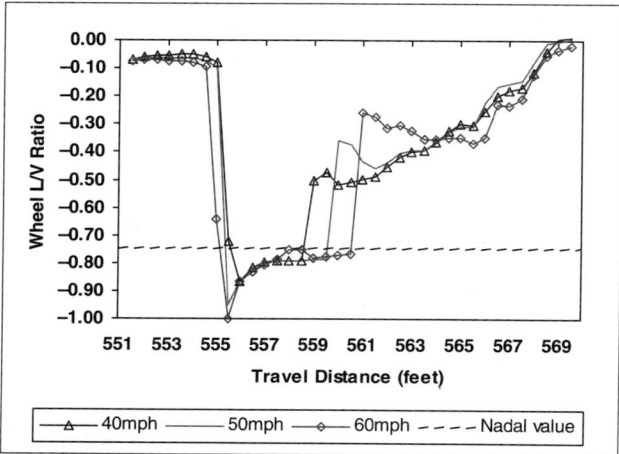

Figure B-47. Wheel L/V ratio at different speeds (Heavy Rail Vehicle, Wheel 1).

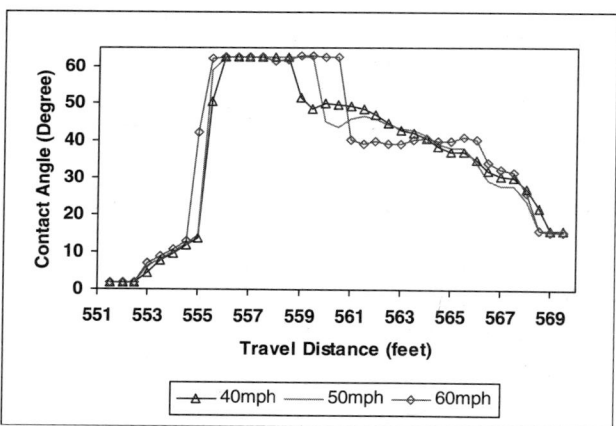

Figure B-48. Wheel contact angle at different speeds (Heavy Rail Vehicle, Wheel 1).

This value is based on the single wheelset simulations presented in Chapter 2, with a flanging wheel friction coefficient of 0.5 and a nonflanging wheel friction coefficient of 0.001. Because the whole vehicle simulations are made with a friction coefficient of 0.5 on both wheels, the criterion given by Equation B-3 is conservative. It is appropriate to use the conservative value because of the uncertainty of actual friction coefficients in service.

When the wheel L/V ratio is over the limiting value, the wheel begins climbing on the flange and the climbing process ends at the point when the L/V ratio is lower than the limiting value. This climb distance calculation method is used for the analysis of all vehicle simulation results in Chapter 3. This is a more conservative and practical definition than used in Chapter 2, which was defined based on the flange angle of 26.6 degrees on the flange tip.

The corresponding climb distances listed in Table B-4 show that the climb distance becomes longer with increasing speed.

The curvature at this track location is 1.95 degrees (see Figure B-41). According to Equation B-4, the climb distance limit is 4.0 ft. When the vehicle travels at speeds lower than 50 mph, the climb distance is less than or equal to the limiting value, the wheel climbs to the maximum flange angle, and the contact angle remains at 63 degrees (Figure B-48). However, when the running speed is increased to 60 mph, the climb distance is 5.6 ft, which is over the limiting value. The wheel climbs above the maximum flange angle and reaches the flange tip between the distances of 557.5 and 559 ft. The contact angle changes from 63 degrees to 61.5 degrees, and the L/V ratio drops.

As seen in Figure B-49, the RRD also increases significantly, indicating that the wheel is climbing the flange and making contact on the flange tip. Although the wheel falls back from flange contact to wheel tread contact after the climb process, it is regarded as an incipient derailment. It is a high risk for a vehicle to run in this situation; a small disturbance during practical running could lead to a derailment.

As shown above, when the climb distance is greater than the proposed criterion value, the wheel climbs over the maximum flange angle onto the flange tip. Derailment could happen due to any small disturbance. Although derailment has not actually occurred, in practical terms it is considered an unsafe condition when the climb distance exceeds the proposed criterion. The vehicle simulation results confirm the methodology and criterion proposed for Wheelset 1 in Section 2.1.

TABLE B-4 First axle wheel-climb distance (ft) (Heavy Rail Vehicle, Wheel 1)

Speed	Start Climbing Point	End Climbing Point	Climb Distance
40 mph	555.5	558.5	3
50 mph	555.5	559.5	4
60 mph	555.2	560.5	5.3

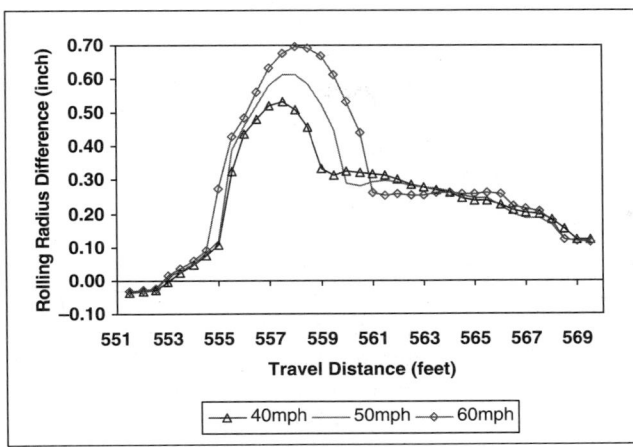

Figure B-49. The wheel RRD at different speeds (Heavy Rail Vehicle, Wheel 1).

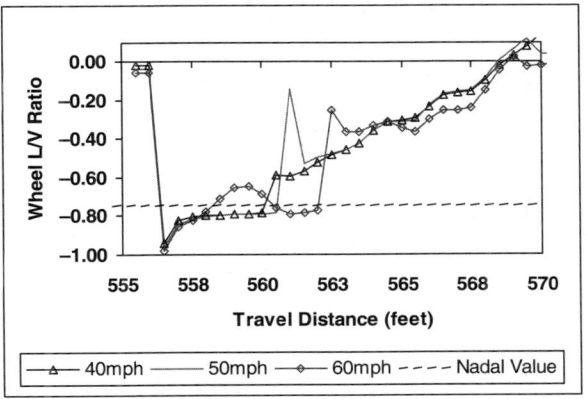

Figure B-50. Wheel L/V ratio at different speeds (Heavy Rail Vehicle, Wheel 2).

3.1.3.2 Heavy Rail Vehicles Assembled with Wheelset 2

Simulations for vehicles assembled with Wheelset 2 also show similar results to Wheelset 1: the climb distance becomes longer with the increasing speed, as shown in Table B-5 and Figure B-50.

The Nadal value for the Wheel 2 profile is 0.745, the proposed L/V ratio limit value is 0.74, and the climb distance limit is 3.1 ft according to Equation B-11.

When the vehicle travels at 50 mph, the climb distance is 4.0 (longer than the limiting value 3.1), the wheel still climbs on the maximum flange angle parts, and the contact angle stays at 63 degrees as seen in Figure B-51. This indicates that the proposed criterion for Wheel 2 is conservative for this situation.

Figure B-53 shows the variation of wheelset AOA during climb. According to the analysis in Chapter 2, the decrease of AOA leads to an increase of climb distance. But the climb distance is 5.6 ft when the running speed is increased to 60 mph, which is above the limiting value. The wheel climbs over the maximum flange angle and onto the flange tip between the distances of 558 and 561 ft. The contact angles changes from 63 degrees to 57 degrees, with a significant drop in the L/V ratio and contact angle. As seen in Figure B-52, the RRD also increases significantly when the wheel contacts at flange tip positions.

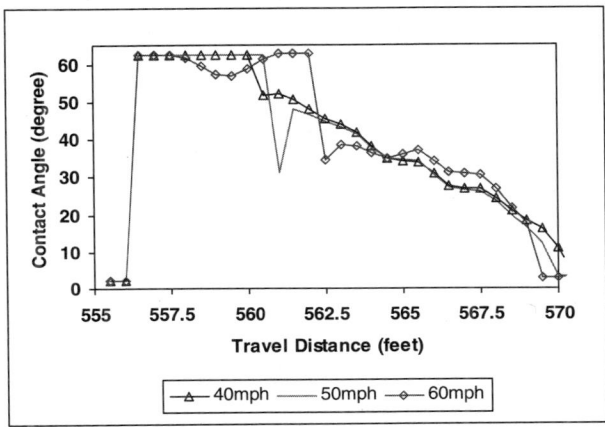

Figure B-51. Wheel contact angle at different speeds (Heavy Rail Vehicle, Wheel 2).

In contrast to Wheel 1, when climbing onto the flange tip occurs, the contact angle of Wheel 2 reduces by 3 degrees while Wheel 1 reduces by only 1.5 degrees. Wheel 2 also climbs even farther onto the tip of the flange. As noted in Table B-3, the flange length of Wheel 2 is 9 mm shorter than that of Wheel 1. Therefore, although both wheels have the same maximum flange angle, the safety margin for Wheel 2 is even smaller, and its derailment probability is significantly increased.

The simulation results for the heavy rail vehicles assembled with two different types of wheel profiles confirm the methodology and criteria proposed for Wheel 1 and 2 in Chapter 2.

TABLE B-5 First axle wheel-climb distance (ft) (Heavy Rail Vehicle, Wheel 2)

Speed (mph)	Start Climbing Point	End Climbing Point	Climb Distance
40	556.4	560	3.6
50	556.4	560.4	4.0
60	556.4	562	5.6

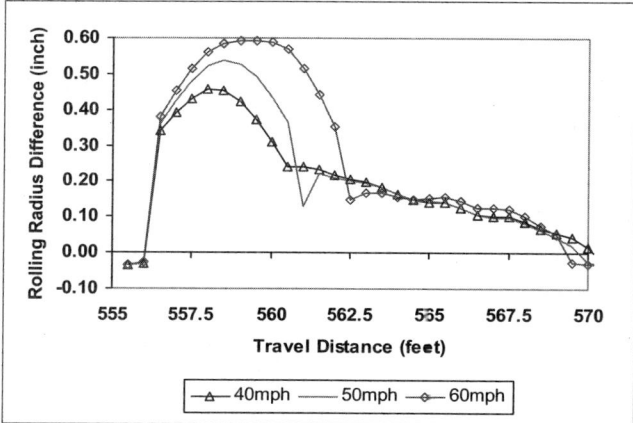

Figure B-52. The wheel RRD at different speeds (Heavy Rail Vehicle, Wheel 2).

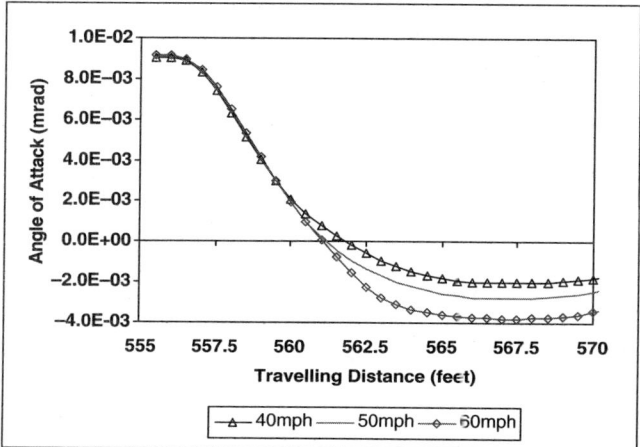

Figure B-53. Wheelset AOA at different speeds (Heavy Rail Vehicle, Wheel 2).

3.2 ARTICULATED LOW FLOOR LIGHT RAIL VEHICLE (MODEL 1)

3.2.1 The Vehicle Model

The vehicle modeled is a typical articulated low floor light rail transit vehicle. It is composed of three car bodies with three trucks. The end car bodies are each mounted on a single truck at one end and connected to an articulation unit at the other end. The center car body is the articulation unit riding on a single truck with independent rotating wheels. The principal dimensions of the vehicle are as follows: (1) rigid wheel base 74.8 in., (2) solid wheel diameter 28 in., (3) independent rotating wheel diameter 26 in., and (4) truck centers 289.4 in. Overall, a total of 26 bodies and 138 connections were used to assemble the simulation model. The rail/wheel friction coefficient is 0.5. The track input model is the same as used for the heavy rail vehicle as described in Section 3.1.2.

3.2.2 Low Floor Light Rail Vehicle (Model 1) Simulation Results

Simulations for the articulated low floor light rail vehicle (Model 1) were made for four different wheel profiles to validate the corresponding proposed L/V ratio and distance-to-climb criteria.

The first wheel/rail profile combination used was Wheel 4 on standard AREMA 115 10-lb/yd rail. This is the same as that used for the Wheel 4 single wheelset simulations discussed in Section 2.4. The rail/wheel friction coefficient is 0.5. This combination was applied to the end trucks of the articulated light rail vehicle.

The second wheel/rail profile combination used was Wheel 5 (IRW) on standard AREMA 115 10-lb/yd rail. This is the same as that used for the Wheel 5 single wheelset simulations discussed in Section 2.5. The rail/wheel friction coefficient is 0.5. This combination was applied to the middle truck on the articulation unit of the light rail vehicle.

Both of these wheel profiles have identical shapes with a 75-degree flange angle. However, Wheel 5 has independent rotating wheels.

The third wheel/rail profile combination used was Wheel 3 on standard AREMA 115 10-lb/yd rail. This is the same as that used for the Wheel 3 single wheelset simulations discussed in Section 2.3. The flange angle is 60 degrees and the rail/wheel friction coefficient is 0.5. This combination was applied to the end trucks of the articulated light rail vehicle.

The fourth combination is the same profile as the third, but modified to have independent rotating wheels for application to the middle truck on the articulation unit of the light rail vehicle.

The simulations were conducted for a range of speeds to generate a range of flange climbing conditions:

- Contact with maximum flange angle, but not flange climbing.
- Flange beginning to climb up the rail, but not derailing (incipient derailment).
- Flange climbing that terminated in derailment.

This range of conditions represents what happens to actual vehicles when they encounter severe track perturbations. The proposed criteria were evaluated by comparing them to the results for these different flange climb conditions.

3.2.2.1 Low Floor Light Rail Vehicle (Model 1) Assembled with Wheelsets 4 and 5

The first simulations of the light rail vehicle Model 1 were conducted with Wheelset 4 on the end trucks and Wheelset 5 (independent rotating wheels) on the center

TABLE B-6 Third axle wheel-climb distance (ft) (Light Rail Vehicle 1, Wheel 5)

Speed (mph)	Start Climbing Point	End Climbing Point	Climb Distance
20	579.8	582.8	3
30	579.8	584.6	4.8
37	578.9	587.4	8.5

truck under the articulation unit. Simulation results show the following:

- The third axle begins to climb near the location of 579 ft distance (distance referred to the third axle).
- The climb distance becomes progressively longer at higher speeds, as shown in Table B-6 and Figure B-54.
- The third wheelset (Wheel 5, IRW) derails at the speed of 39 mph.

For Wheel 5 profile, the Nadal limiting L/V ratio is 1.13. According to Equation B-20, the proposed L/V ratio limiting value is 1.0. The curvature at this location is 1.95 degrees and the climb distance limit is 5.2 ft, according to Equation B-23.

When the vehicle travels at speeds lower than 30 mph, the flange-climb distance is less than the limiting value. The wheel climbs onto the maximum flange angle and the contact angle stays at 75 degrees, as seen in Figure B-55. However, when the running speed is increased to 37 mph, the climb distance increases to 8.5 ft, which is over the limit value, and the wheel climbs over the maximum flange angle and reaches the flange tip between the distances of 581.7 and 582.3 ft. The contact angle changes from 74.5 degrees to 73.9 degrees, with an insignificant drop in the L/V ratio. However, the RRD (shown in Figure B-56) increases significantly when the wheel contacts at the flange tip position, clearly showing that the wheel has climbed over the maximum flange angle and is running on the flange tip.

As shown in Figures B-54 through B-56, the third axle derails when the running speed is further increased to 39 mph.

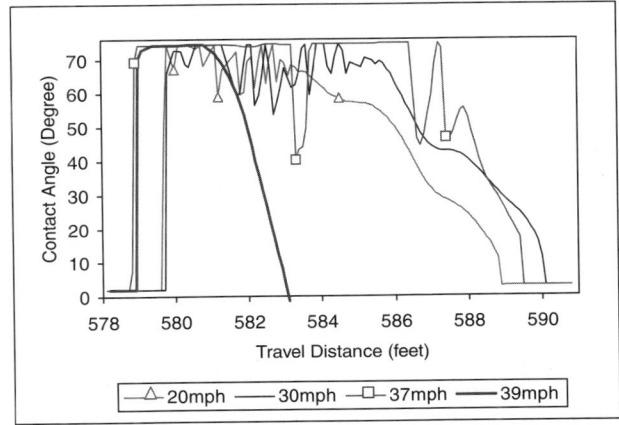

Figure B-55. Wheel contact angle at different speeds (Light Rail Vehicle 1, Wheel 5).

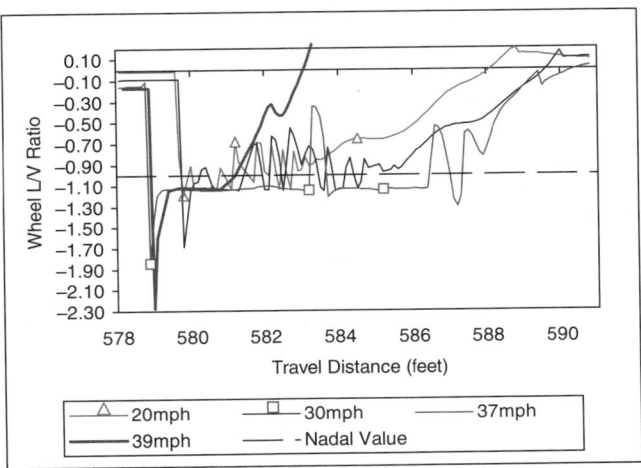

Figure B-54. Wheel L/V ratio at different speeds (Light Rail Vehicle 1, Wheel 5).

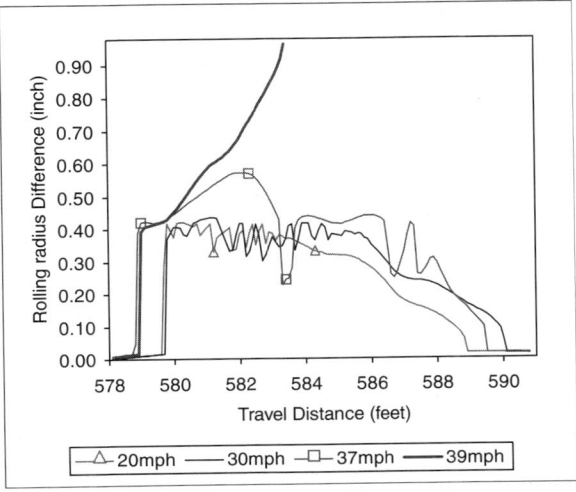

Figure B-56. Wheel RRD at different speeds (Light Rail Vehicle 1, Wheel 5).

TABLE B-7 Third axle wheel-climb distance (ft) (Light Rail Vehicle 1, IRW Wheel 3)

Speed (mph)	Start Climbing Point	End Climbing Point	Climb Distance
7	580.1	581.9	1.8
10	580.2	583.5	3.3
11	580.2	581.5	1.3, Derail

This result gives a better demonstration of the high risk for vehicles to run when the climb distance is over the limiting value. The simulation results also confirm the validation of the proposed climb distance criteria discussed in Chapter 2.

3.2.2.2 Low Floor Light Rail Vehicles Assembled with Solid and IRW Wheelset 3

The second set of simulations of the light rail vehicle Model 1 were conducted with Wheelset 3 on the end trucks and Wheelset 3 modified with independent rotating wheels on the center truck under the articulation unit. Simulation results show the following:

- The third axle begins to climb at the location of 580 ft distance (distance referred to the third axle).
- The climb distance becomes longer with increasing speed before derailment occurs, as shown in Table B-7 and Figure B-57.
- The third wheelset (independent rotating wheels) derails at a speed of 11 mph.

For the Wheel 3 profile, the Nadal value is 0.67. According to Equation B-14, the proposed L/V ratio limiting value is 0.66. The curvature at this location is 1.95 degrees and the climb distance limit is 3.3 ft according to Equation B-15.

When the vehicle travels at speed lower than 10 mph, the climb distance is less than or equal to the limiting value. The wheel climbs onto maximum flange angle and the contact angle stays at 60 degrees, as seen in Figure B-58.

However, when the vehicle travels at a speed of 10 mph, the climb distance equals the limiting value. Although the wheel still stays at the maximum flange angle, it has climbed farther up the flange than at 7 mph, as shown by the RRD in Figure B-59.

When the running speed is increased a little more to 11 mph, the wheel climbs above the maximum flange angle, over the flange tip, and ultimately derails. In contrast to Wheel 5 discussed in the previous section, the IRW Wheel 3 is unacceptable for this kind of light rail vehicle. When derailment occurred, the climb distance was very rapid and much shorter than the proposed limiting value, which is to be expected.

Compared to Wheel 5, the RRD for IRW Wheel 3 in Figure B-59 increases significantly even when the wheel is still climbing on the maximum flange angle. This is an important characteristic for low maximum flange angle profile wheels. In other words, the very low maximum flange angle makes it easy for the wheelset to climb up on the flange tip,

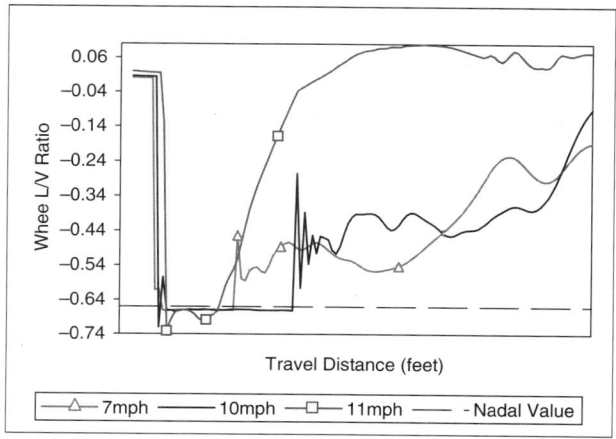

Figure B-57. Wheel L/V ratio at different speeds (Light Rail Vehicle 1, Wheel 3).

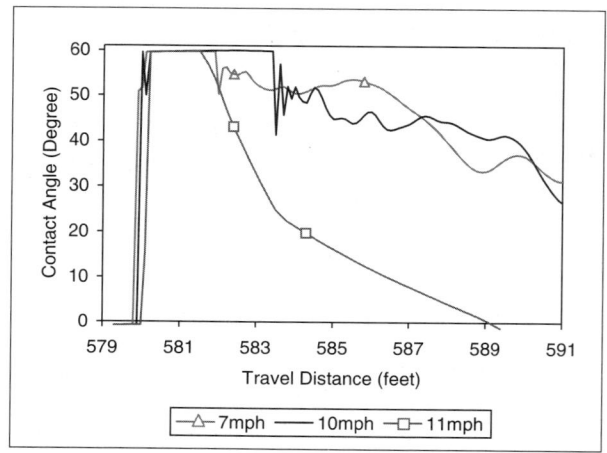

Figure B-58. Wheel contact angle at different speeds (Light Rail Vehicle 1, Wheel 3).

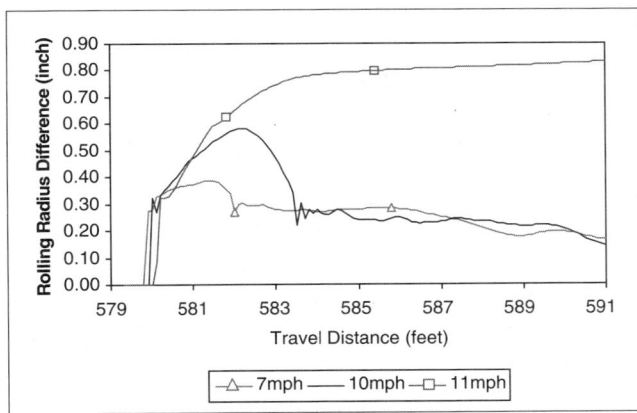

Figure B-59. Wheel RRD at different speeds (Light Rail Vehicle 1, Wheel 3).

even though there may be a longer maximum flange angle length.

The simulation results for light rail vehicles assembled with Wheel 5 and Wheel 3 IRW profile wheelsets also confirm the methodology and criteria proposed for them in Chapter 2.

The significant difference in the simulation results for these two wheel profiles shows that optimization of wheel profiles are extremely important in the design of a particular vehicle.

3.3 ARTICULATED HIGH FLOOR LIGHT RAIL VEHICLE (MODEL 2)

3.3.1 The Vehicle Model

The articulated high floor light rail vehicle Model 2, composed of two car bodies and three trucks, represents another typical type of articulated transit system vehicle. The two car bodies articulate on the middle truck, with all three trucks having solid wheelsets. The principal dimensions of the vehicle are as follows: (1) rigid wheel base 75 in., (2) wheel diameter 26 in., and (3) truck centers 275.5 in. Overall, a total of 18 bodies and 85 connections were used to assemble the simulation model. The rail/wheel friction coefficient is 0.5. The track input model is the same as described in Section 3.1.2.

3.3.2 High Floor Light Rail Vehicle (Model 2) Simulation Results

Simulations for the high floor articulated light rail vehicle ratio (Model 2) were made for two different wheel profiles to validate the corresponding proposed L/V ratio and distance-to-climb criteria.

The first wheel/rail profile combination used was Wheel 2 on standard AREMA 115 10-lb/yd rail. This combination is the same as that used for the Wheel 2 single wheelset simulations discussed in Section 2.2. The rail/wheel friction coefficient is 0.5. The flange angle is 63 degrees.

The second wheel/rail profile combination used was Wheel 3 on standard AREMA 115 10-lb/yd rail. This is the same as that used for the Wheel 3 single wheelset simulations discussed in Section 2.3. The rail/wheel friction coefficient is 0.5. The flange angle is 60 degrees.

The simulations were conducted for a range of speeds to generate a range of flange climbing conditions:

- Contact with maximum flange angle, but not flange climbing.
- Flange beginning to climb up the rail, but not derailing (incipient derailment).
- Flange climbing that terminated in derailment.

This range of conditions represents what happens to actual vehicles when they encounter severe track perturbations. The proposed criteria were evaluated by comparing them to the results for these different flange climb conditions.

3.3.2.1 High Floor Light Rail Vehicle (Model 2) Assembled with Wheelset 2

Simulation results show the following:

- The first axle begins to climb near the location of 555 ft distance (distance referred to the third axle).
- The climb distance becomes progressively longer with increasing speed, as shown in Table B-8 and Figure B-60.

TABLE B-8 Third axle wheel-climb distance (ft) (Light Rail Vehicle 2, Wheel 2)

Speed (mph)	Start Climbing Point	End Climbing Point	Climb Distance
20	555.2	558.5	3.2
30	555.1	559	3.9
40	555	560	5

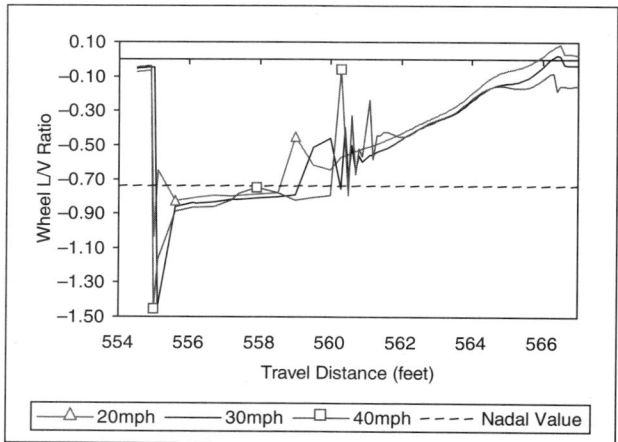

Figure B-60. Wheel L/V ratio at different speeds (Light Rail Vehicle 2, Wheel 2).

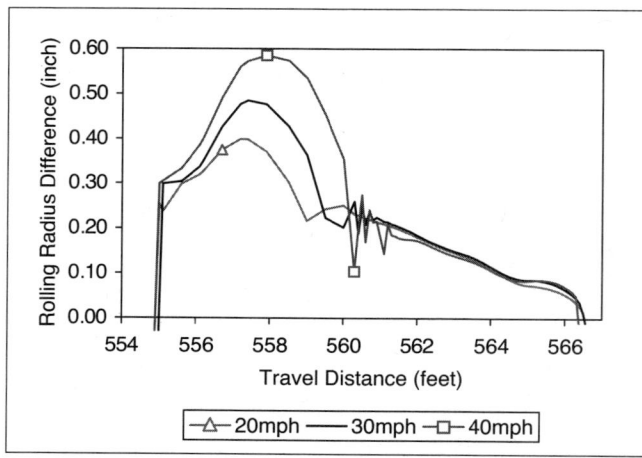

Figure B-62. Wheel RRD at different speeds (Light Rail Vehicle 2, Wheel 2).

The Nadal value and proposed criterion for Wheel 2 have been listed in Section 3.1.3.2. The proposed L/V ratio limiting value is 0.74, and the proposed flange-climb-distance limit is 3.1 ft.

When the vehicle travels at speeds lower than 30 mph, the climb distance is longer than the limiting value. The wheel climbs to the maximum flange angle face, and the contact angle stays at 63 degrees, as seen in Figure B-61. The proposed criterion for Wheel 2 is conservative for this situation (the same conclusion has been found in 3.1.3.2).

However, when the running speed is increased to 40 mph, the climb distance is 5 ft—much higher than the limiting value. The wheel climbs above the maximum flange angle and onto the flange tip between the distances of 557.2 and 557.9 ft. The contact angle reduces from 63 degrees to 59.5 degrees, with a significant drop in the L/V ratio. At the same time, the RRD (shown in Figure B-62) increases significantly when the wheel contacts the rail on the flange tip. The vehicle is running unsafely in this condition.

The simulation results of light rail vehicles (Model 2) assembled with Wheel 2 show that the criterion proposed for Wheel 2 in Section 2 is conservative at low speed, which is consistent with the conclusion in Section 3.1.1.2 for heavy rail vehicles assembled with the same profile wheelsets. However, the proposed criterion is still valid when the climb distance is much higher than the limit.

3.3.2.2 High Floor Light Rail Vehicle (Model 2) Assembled with Wheelset 3

The Nadal value and proposed criterion for Wheel 3 have been listed in Section 3.2.2.2. The proposed L/V ratio limit value is 0.66, the proposed climb distance limit is 3.3 ft.

The simulation results for light rail vehicles (Model 2) assembled with Wheel 3 are shown in Table B-9 and Figures B-63 through B-65. In general the results are similar to those of Wheel 2.

The simulation results also show that the proposed climb distance criterion for Wheel 3 is valid when the climb distance is very much over the limit value, although it is conservative for low speed situations.

In contrast to Wheel 2, Wheel 3 takes more distance to climb at the same speed even though it has a smaller flange angle. This is because Wheel 3 has a longer flange length, which allows the wheel to climb in a longer distance.

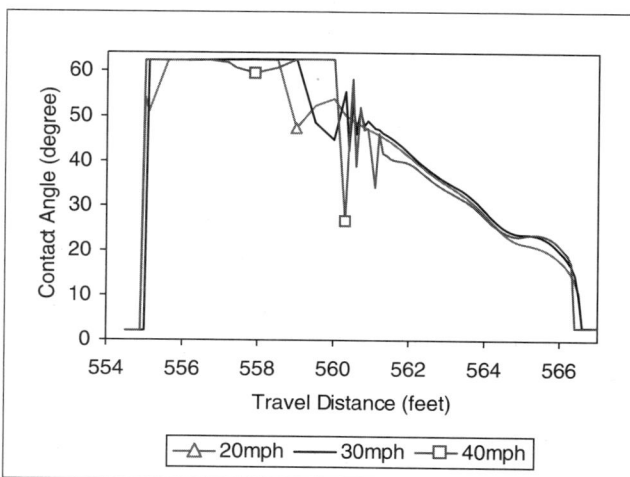

Figure B-61. Wheel contact angle at different speeds (Light Rail Vehicle 2, Wheel 2).

TABLE B-9 Third axle wheel-climb distance (ft) (Light Rail Vehicle 2, Wheel 3)

Speed (mph)	Start Climbing Point	End Climbing Point	Climb Distance
15	555.5	559	3.5
20	555.3	559.3	4.0
35	555.1	560.7	5.6

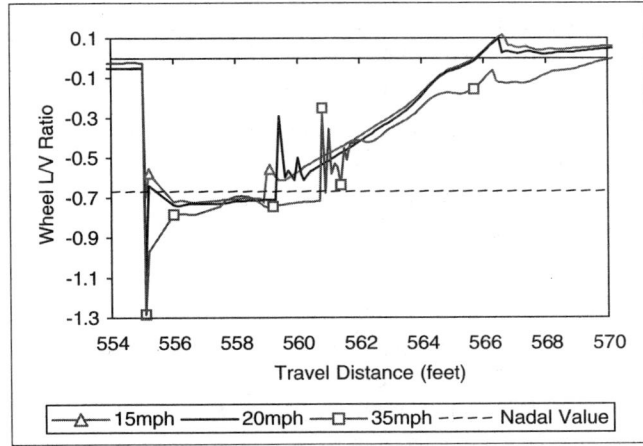

Figure B-63. Wheel L/V ratio at different speeds (Light Rail Vehicle 2, Wheel 3).

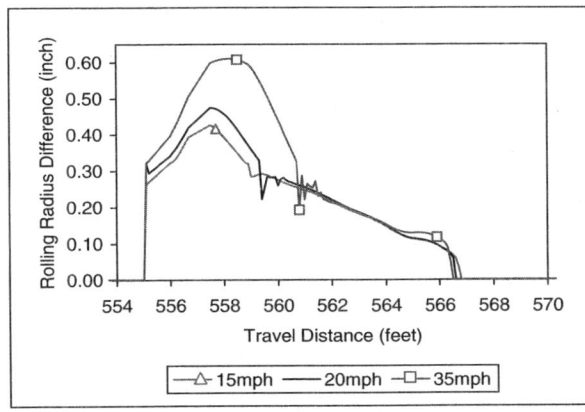

Figure B-65. Wheel RRD at different speeds (Light Rail Vehicle 2, Wheel 3).

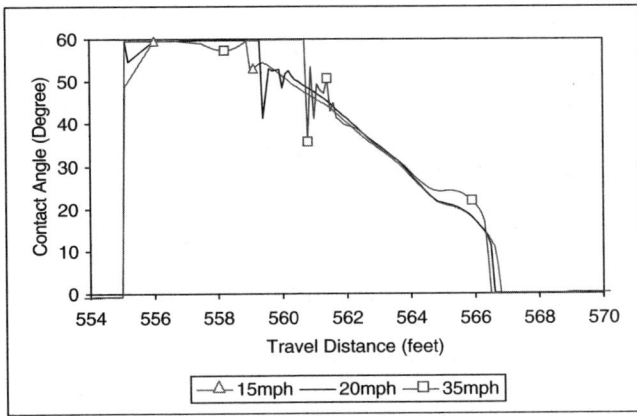

Figure B-64. Wheel contact angle at different speeds (Light Rail Vehicle 2, Wheel 3).

3.4 VEHICLE SIMULATION SUMMARY

In general, the simulation results for these three types of transit vehicles assembled with five different types of wheel profiles confirm the validity of the methodology and criteria proposed in Chapter 2. The incipient derailment can be predicted by applying these criteria in vehicle dynamics simulation analysis.

The simulation results also show that the proposed climb distance criteria for low-maximum-flange-angle wheelsets are conservative at low speeds. For the simulations shown, once the flange climb reached the maximum flange angle the AOA began to reduce for two reasons:

- Increased rolling radius causes the wheelset to start steering back (this does not happen for the IRW).
- The track perturbation geometry changes, reducing the AOA.

This reduction in AOA increases the effective L/V ratio limit and lengthens the effective flange-climb-distance limit. This has the effect, in general, of making the proposed criteria conservative.

Because the L/V ratios and climb distance are sensitive to the wheelset AOA, the effects of AOA variation during climb need to be further investigated both by single wheelsets and vehicles simulations.

CHAPTER 4

CONCLUSIONS AND DISCUSSIONS

4.1 CONCLUSIONS

Based on the single wheelset and vehicle simulation results, the following conclusions are drawn:

- New single wheel L/V distance criteria have been proposed for transit vehicles with specified wheel profiles:

Wheel 1 profile:

$$\text{L/V Distance (feet)} < \frac{5}{0.13 * AOA + 1},$$
$$\text{if } AOA < 10 \text{ mrad}$$

$$\text{L/V Distance (feet)} = 2.2, \quad \text{if } AOA \geq 10 \text{ mrad}$$

Wheel 2 profile:

$$\text{L/V Distance (feet)} < \frac{4.1}{0.16 * AOA + 1},$$
$$\text{if } AOA < 10 \text{ mrad}$$

$$\text{L/V Distance (feet)} = 1.6, \quad \text{if } AOA \geq 10 \text{ mrad}$$

Wheel 3 profile:

$$\text{L/V Distance (feet)} < \frac{4.2}{0.136 * AOA + 1},$$
$$\text{if } AOA < 10 \text{ mrad}$$

$$\text{L/V Distance (feet)} = 1.8, \quad \text{if } AOA \geq 10 \text{ mrad}$$

Wheel 4/5 profile:

$$\text{L/V Distance (feet)} < \frac{28}{2 * AOA + 1.5},$$
$$\text{if } AOA < 10 \text{ mrad}$$

$$\text{L/V Distance (feet)} = 1.3, \quad \text{if } AOA \geq 10 \text{ mrad}$$

Wheel 6 profile:

$$\text{L/V Distance (feet)} < \frac{49}{2 * AOA + 2.2},$$
$$\text{if } AOA < 10 \text{ mrad}$$

$$\text{L/V Distance (feet)} = 2.2, \quad \text{if } AOA \geq 10 \text{ mrad}$$

where AOA is in mrad. In situations where AOA is not known and cannot be measured, the equivalent AOA (AOAe) calculated from curve curvature and truck geometry should be used in the above criteria.

- If AOA is known and can be measured, more accurate new single wheel L/V ratio criteria based on AOA have also been proposed (see corresponding equation in Chapter 2).
- The simulation results for transit vehicles assembled with different types of wheel profiles confirm the validity of the proposed criteria.
- For most conditions, an incipient derailment occurs when the climb distance exceeds the proposed criterion value.
- The proposed climb distance criteria are conservative for most conditions. Under many conditions, variations of AOA act to reduce the likelihood of flange climb.
- The single wheel L/V ratio required for flange climb derailment is determined by the wheel maximum flange angle, friction coefficient, and wheelset AOA.
- The L/V ratio required for flange climb converges to Nadal's value at higher AOAs (above 10 mrad). For lower wheelset AOAs, the wheel L/V ratio necessary for flange climb becomes progressively higher than Nadal's value.
- The distance required for flange climb derailment is determined by the L/V ratio, wheel maximum flange angle, wheel flange length, and wheelset AOA.
- The flange-climb distance converges to a limiting value at higher AOAs and higher L/V ratios. This limiting value is highly correlated with wheel flange length. The longer the flange length, the longer the climb distance. For lower wheelset AOAs, when the L/V ratio is high enough for the wheel to climb, the wheel-climb distance for derailment becomes progressively longer than the proposed flange-climb-distance limit. The wheel-climb distance at lower wheelset AOA is mainly determined by the maximum flange angle and L/V ratio.
- Besides the flange contact angle, flange length also plays an important role in preventing derailment. The climb distance can be increased through use of higher wheel maximum flange angles and longer flange length.
- The flanging wheel friction coefficient significantly affects the wheel L/V ratio required for flange climb; the lower the friction coefficient, the higher the single wheel L/V ratio required.
- For conventional solid wheelsets, a low nonflanging wheel friction coefficient has a tendency to cause flange

climb at a lower flanging wheel L/V ratio, and flange climb occurs over a shorter distance for the same flanging wheel L/V ratio.
- The proposed L/V ratio and flange-climb-distance criteria are conservative because they are based on an assumption of a low nonflanging wheel friction coefficient.
- For independent rotating wheelsets, the effect of the nonflanging wheel friction coefficient is negligible because the longitudinal creep force vanishes.
- The proposed L/V ratio and flange-climb-distance criteria are less conservative for independently rotating wheels because they do not generate significant longitudinal creep forces.
- For the range of track lateral stiffness normally present in actual track, the wheel-climb distance is not likely to be significantly affected by variations in the track lateral stiffness.
- The effect of inertial parameters on the wheel-climb distance is negligible at low speeds.
- At high speeds, the climb distance increases with increasing wheelset rotating inertia. However, the effect of inertial parameters is not significant at low nonflanging wheel friction coefficients.
- Increasing vehicle speed increases the distance to climb.

4.2 DISCUSSION

An AOA measurement is not usually available in practice. Therefore, the proposed climb distance criteria based on curvature will normally be used. For the situation of a vehicle running on tangent track, the equivalent AOAe is zero because the tangent line curvature is zero. However, under certain track perturbations and running speeds, the wheelset AOA could in practice be very large for some poor-steering trucks, such as typical freight car trucks, very worn passenger trucks, trucks with axle misalignments, and trucks with large turning resistance. Although certain types of trucks (H-frame passenger car trucks, trucks with soft primary suspension) could have small AOAs due to a better steering ability, the criteria must be conservative enough to identify potential bad performance. For the cases on tangent lines, the criteria based on a zero AOAe may not be conservative enough to capture bad trucks.

Most passenger rail cars (including transit and intercity cars) have truck designs that control axle yaw angles better than standard freight cars; and, in some instances, passenger cars have softer primary suspensions that provide for better axle steering, resulting in lower AOAs and longer flange climb distances. Therefore, the proposed criteria for transit cars are made less conservative than freight cars. However, there is no guarantee that all rail passenger cars have better truck designs, and the criteria must be made sufficiently conservative to capture poor performance either from poor track quality or from poor axle steering. Rail passenger cars with good truck designs and good axle steering will meet the more conservative criteria because of their better steering capabilities. The more conservative criteria will provide a greater margin of safety for the better performing vehicles and ensure that the poor performing trucks are captured.

Based on the single wheelset and complete car simulation results, both the L/V ratio and climb distance converge to corresponding limit values when the wheelset AOA is over 10 mrad. Therefore, the 10-mrad AOA situation represents the most conservative case for wheelset climb derailment, which could be used as an alternative criterion for both tangent and curved track line cases together with the proposed criterion in this report. This has the significant advantage of proposing only one L/V criterion and one distance-to-climb criterion for a particular wheel/rail profile combination and they are not dependent on knowing AOA, curvature, truck design, or track perturbation conditions. The resulting criteria for Wheelset 1 using the AOA of 10 mrad would be:

$$\frac{L}{V} < 0.74, \qquad \text{if AOA} \geq 10 \text{ mrad}$$

$$\text{L/V Distance (feet)} < 2.2, \qquad \text{if AOA} \geq 10 \text{ mrad}$$

Although onboard AOA measurement is not available in practice, the wheelset AOA at a specific location can be measured by a wayside measurement system. This system makes the 10-mrad criteria operational in practical running and tests. The 10-mrad criteria need further investigation and evaluation in comparison to the criteria proposed thus far.

A significant concern with the proposed criteria is that they are specific to the particular wheel/rail profile combinations that were analyzed. The criteria appear to be dependent on details of the particular wheel and rail profile shapes. Although similar analyses could be performed to develop new criteria for a specific wheel and rail profile pair, it is recognized that the transit industry would prefer to have some general formulas for calculating flange climb safety criteria for any conditions.

Another concern is that the proposed criteria have been developed based on some simple assumptions of likely wheelset AOA in curved and straight track. Uncertainties regarding differences in the axle steering characteristics of different vehicles and the likelihood of encountering track geometry deviations that can cause local increases in wheelset AOA require that conservative assumptions be made, resulting in proposed criteria that may be too conservative.

The friction coefficient varies with the rail and wheel surface conditions and has important effects on derailment. A climb distance criterion taking the variation of friction coefficients into account will provide more valuable information for wheel/rail interaction mechanisms and rail vehicle safety.

Under the conditions of flange climb, large lateral forces are likely to be present that may cause the rail to roll—especially if the track structure is weak. Rail roll will change the wheel/rail contact conditions and may result in lower effective contact angles and shorter effective maximum flange angle lengths, with consequent reductions in L/V limits and flange climb distances.

The following are specific recommendations for work in the future to complete the validation efforts and to address some of these concerns:

- Perform comparisons with results from full-scale tests to further validate the criteria proposed for transit vehicles.
- Since the climb distance limit is highly correlated with the flange parameters (flange angle, length, height), further investigate and propose a general climb distance criterion that depends on both the AOA and flange parameters.
- Because the L/V ratios and the climb distance are sensitive to the wheelset AOA, further investigate the effect of variations of AOA during flange climb using simulations of both single wheelsets and full vehicles.
- Further develop flange-climb-distance criteria to account for the effects of carrying friction coefficient.
- Perform additional single wheel simulations to investigate the effects of rail rotation.

Because of the complexity of derailments and due to limited funding, only a few of these tasks can be accomplished in Phase II. The rest of the recommended work may need continuing efforts in the future.

APPENDIX B-1:
LITERATURE REVIEW

B1.1 INTRODUCTION

The research work performed for this project was based on methods developed by the research team during tests and analyses performed from 1994 to 1999 (*1, 2*). In recent years, other organizations have also been performing flange climb derailment research. A literature review was conducted to ensure their findings were understood prior to performing this research project.

B1.2 BLADER (*9*)

F. B. Blader (*9*) has given a clear description and discussion of wheel-climb research and safety criteria that had been examined or adopted by railroad operators and railroad test facilities as guidelines for safety certification testing of railway vehicles. Briefly, they are the following:

- Nadal's Single-Wheel L/V Limit Criterion.
- Japanese National Railways' (JNR) L/V Time Duration Criterion.
- GM Electromotive Division's (EMD) L/V Time Duration Criterion.
- Weinstock's Axle-Sum L/V Limit Criterion.

The Nadal single-wheel L/V limit criterion, proposed by Nadal in 1908 for the French Railways, has been used throughout the railroad community. Nadal established the original formulation for limiting the L/V ratio in order to minimize the risk of derailment. He assumed that the wheel was initially in two-point contact with the flange contact point leading the tread, and he concluded that the wheel material at the flange contact point was moving downwards relative to the rail material, due to the wheel rolling about the tread contact point. Nadal further theorized that wheel climb occurs when the downward motion ceases with the friction saturated at the contact point. Based on his assumption and a simple equilibrium of the forces between a wheel and rail at the single point of flange contact, Nadal proposed a limiting criterion as a ratio of L/V forces:

$$\frac{L}{V} = \frac{\tan(\delta) - \mu}{1 + \mu \tan(\delta)}$$

The expression for the L/V ratio criterion is dependent on the flange angle δ and friction coefficient μ. Figure B1-1 shows the solution of this expression for a range of values appropriate to normal railroad operations. The AAR has based its Chapter XI single-wheel L/V ratio criterion on Nadal's theory using a friction coefficient of 0.5.

Following several laboratory experiments and observations of actual values of L/V ratios greater than the Nadal criterion at incipient derailment, researchers at the Japanese National Railways (JNR) proposed a modification to Nadal's criterion (*4*). For time durations of less than 0.05 s, such as might be expected during flange impacts due to hunting, an increase was given to the value of the Nadal L/V criterion. However, small-scale tests conducted at Princeton University indicated that the JNR criterion was unable to predict incipient wheel-climb derailment under a number of test conditions.

A less conservative adaptation of the JNR criterion was used by the Electromotive Division of General Motors (EMD) in its locomotive research (*5*).

More recently, Weinstock of the United States Volpe National Transportation Systems Center observed that this balance of forces does not depend on the flanging wheel alone (*6*). Therefore, he proposed a limit criterion that utilizes the sum of the absolute value of the L/V ratios seen by two wheels of an axle, known as the "Axle Sum L/V" ratio.

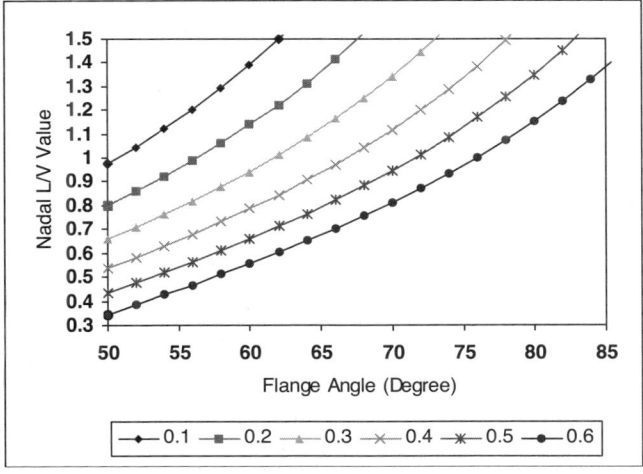

Figure B1-1. Nadal criterion values.

He proposed that this sum be limited by the sum of the Nadal limit (for the flanging wheel) and the coefficient of friction (at the nonflanging wheel). Weinstock's criterion was argued to be not as overly conservative as Nadal's at small or negative AOA and less sensitive to variations in the coefficient of friction.

The Weinstock criterion retains the advantage of simplicity. It can be measured with an instrumented wheelset, which measures the values of L/V ratio on both wheels. It is not only more accurate than Nadal's criterion, it also has the merit of being less sensitive to errors or variations in the coefficient of friction.

B1.3 50-MS DISTANCE CRITERION

Based on the JNR and EMD research, and considerable experience in on-track testing of freight vehicles, a 0.05-s (50-ms) time duration was adopted by the AAR for the Chapter XI certification testing of new freight vehicles. This time duration has since been widely adopted by test engineers throughout North America for both freight and passenger vehicles.

B1.4 FRA TRACK SAFETY STANDARDS (7)

A flange-climb-distance limit of 5 ft was adopted by the FRA for their Class 6 high speed track standards (7). This distance limit appears to have been based partly on the results of the joint AAR/FRA flange climb research conducted by TTCI and also on experience gained during the testing of various commuter rail and long distance passenger vehicles.

B1.5 PREVIOUS TTCI RESEARCH (1–3)

The results of wheel climb (also called flange climb) derailment testing and mathematical simulations performed with the TLV at the TTCI are summarized in Reference 1. The important conclusions are the following:

- Large flanging rail friction and nonflanging friction during the test resulted in axle sum L/V ratios at wheel climb that were lower than the Chapter XI limit of 1.5.
- All L/V force ratios found in the tests and with NUCARS simulations converged to the Nadal and Weinstock values at higher AOAs (10 to 15 mrad). At lower and negative AOAs, the predicated and measured L/V ratios exceeded Nadal and Weinstock values.
- The wheel/rail coefficient of friction, the maximum wheel/rail contact angle, and the wheelset AOA have a major influence on the potential for wheel climb.
- Vertical load unbalance does not affect the critical L/V values as computed by the Weinstock and Nadal equations and the L/V ratios measured experimentally.

Nadal's original formulation assumes the worst scenario—that of a zero longitudinal creepage between wheel and rail. Shust et al. (1) propose a modified formulation using the effective coefficient of friction to replace the friction coefficient in Nadal's formulation. The modified formulation is considered less conservative as it accounts for the presence of longitudinal creep forces that tend to provide a stabilizing effect to the wheel climb.

Following the extensive tests of Reference 1, TTCI performed theoretical simulations of flange climb using the NUCARS model. This resulted in proposing a new maximum L/V ratio limit and flange-climb-distance limit (2). These were subsequently revised and presented in Reference 3.

B1.6 DYNAMIC SAFETY (DYSAF) RESEARCH (10, 11)

Kik et al.'s "Comparison of Result of Calculations and Measurements of DYSAF-tests" (10) compares results of calculations and measurements from this research project: DYSAF (assessment of vehicle-track interaction with special reference to DYnamic SAFety in operating conditions). A test running gear was developed to test derailment of a wheelset in guiding and unloaded conditions. The aim of this project was to investigate safety limits of derailment at high speed. The test was carried out in Velim, Czech Republic, in August 2000 to analyze derailment conditions at higher speeds up to 160 km/h. The test was performed in quasi-stationary conditions on a small circuit at low speeds (from 20 to 75 km/h) and dynamic conditions on a great circuit at high speed (from 80 to 160 km/h). The influence of AOA L/V ratio and duration of L/V were investigated in different test series. An extension to the existing Nadal's formula was developed, but parameter identification in this formula has not been done yet.

The lateral, vertical, and roll movement of rails and lateral movement of sleepers were included in the simulation model of 21 rigid bodies with 93 degrees of freedom (DOF). Measured track irregularities, including gage as well as lateral and vertical alignment of left and right rail, were also studied in the simulation. Special effort was made in the identification of simulation parameters such as friction coefficients. The authors reached the following conclusions:

- For the higher velocity, the L/V ratio is much more dominated by higher frequency dynamics and it can no longer be neglected. Measurement of track irregularities should be improved to include the smaller wavelength defects.
- A reasonable threshold of L/V ratio as a derailment criterion or a general multicriterion based on L/V ratio could not be derived until now. Simulation might be the best solution for safety investigations of railway vehicles.

Results of single wheelset derailment simulations, conducted as a part of the DYSAF research project, are presented by Parena et al. in "Derailment Simulation, Parametric Study" (*11*). The simulation cases were based on a wheelset model forced to derail by a lateral force on the level of track with and without excitations in vertical and roll direction, and excitations in the lateral direction. The influence of rail/wheel geometric and friction parameters, vertical loading, and lateral loading duration was investigated. The following conclusions were reached:

- Without influence of lateral sliding, a revised Nadal formula with 3/4 friction coefficient is quite useful to compute maximum L/V ratio.
- Maximum L/V ratio occurs higher up the flange than the maximum angle of the flange, seemingly due to the lateral sliding of the wheel on the rail.
- Nonsymmetric, low frequency vertical loading or lateral force excitation of longer duration reduces the maximum lateral force that the wheelset resists until derailment.
- In any case, bounce and the lower sway of a vehicle should have different Eigen frequencies. If they are excited with nearly the same frequency, only very low lateral force might let the vehicle derail.

B1.7 CLEMENTSON AND EVANS (*12*)

Two real derailment incidents were investigated by Clementson and Evans (*12*). The first case study concerned the derailment of a loaded train of two-axle coal hopper cars on straight track. This derailment was caused by a combination of cyclic twist and lateral and vertical alignment in the rails causing rocking of the cars. Dynamic simulations showed the build up of a swaying motion in the vehicle and the wheels lifting substantially off the rails at the point of derailment. The body roll and wheel loads confirmed a rolling response to the track geometry that resulted in the cyclic unloading of the wheels. At the derailment speed, it was found that the wheelsets were hunting.

The second case study concerned the derailment by flange climbing of a loaded steel coil-carrying car fitted with Y25 bogies. Dynamic simulations showed that unequal dips in the two rails caused a pitching and swaying response of the wagon, which unloaded the leading outer wheel just as it ran into a lateral misalignment giving rise to a very high L/V ratio and subsequent flange climbing. An additional contributory factor was a fault in the vehicle suspension giving rise to an unequal static load distribution across the leading wheelset, combined with offset loading of the steel coil above the leading bogie. Simulations were carried out at 40, 45, and 50 mph with three different load conditions. For nominal vehicles, the L/V ratio increased and was sustained for a longer distance. As the speed increased, the flange climbed to 3 mm and then dropped back. For the asymmetric vehicle, the flange climbed 22 mm to the flange tip and then derailed.

B1.8 CHEN AND JIN (*13*)

In "On a New Method for Evaluation of Wheel Climb Derailment," Chen and Jin (*13*) propose a derailment index for evaluation of the wheel-climb derailment with the measurement of primary suspension forces. The purpose of the adoption of primary suspension forces was to replace the quasi-steady wheel/rail contact forces with dynamic suspension forces for the calculation of derailment index. The derailment index was dependent on the wheelset AOA and vertical unloading ratio.

B1.9 LITERATURE REVIEW SUMMARY

Although considerable research into flange climb is underway, there were no new criteria proposed. The only new criteria, single wheel L/V ratio criterion and L/V distance criterion for freight cars, were proposed by Wu and Elkins (*2*) and revised by Elkins and Wu (*3*). These were developed through wheel/rail interaction analysis and extensive NUCARS simulations. The criteria are strongly dependent on AOA. If AOA cannot be measured, a reduced limit depending on curvature is recommended.

REFERENCES

1. Shust, W.C., Elkins, J., Kalay, S., and El-Sibaie, M., "Wheel-Climb Derailment Tests Using AAR's Track Loading Vehicle," AAR report R-910, Association of American Railroads, Washington, D.C., December 1997.
2. Wu, H., and Elkins, J., "Investigation of Wheel Flange Climb Derailment Criteria," AAR report R-931, Association of American Railroads, Washington, D.C., July 1999.
3. Elkins, J., and Wu, H., "New Criteria for Flange Climb Derailment," IEEE/ASME Joint Railroad Conference paper, Newark, New Jersey, April 4-6, 2000.
4. Matsudaira, T., "Dynamics of High Speed Rolling Stock," Japanese National Railways RTRI Quarterly Reports, Special Issue, 1963.
5. Koci, H.H., and Swenson, C.A., "Locomotive Wheel-Loading—A System Approach," General Motors Electromotive Division, LaGrange, IL, February 1978.
6. Weinstock, H., "Wheel Climb Derailment Criteria for Evaluation of Rail Vehicle Safety," Proceedings, ASME Winter Annual Meeting, 84-WA/RT-1, New Orleans, Louisiana, 1984.
7. Federal Railroad Administration, *Track Safety Standards*, Part 213, Subpart G, November 1998.
8. Wu, H., Shust, W.C., and Wilson, N.G., "Effect of Wheel/Rail Profiles and Wheel/Rail Interaction on System Performance and Maintenance in Transit Phase I Report," Transit Cooperative Research Program report, February 2004.
9. Blader, F.B, "A Review of Literature and Methodologies in the Study of Derailments Caused by Excessive Forces at the Wheel/Rail Interface," AAR report R-717, Association of American Railroads, Washington, D.C., December 1990.
10. Kik, W., et al., "Comparison of Result of Calculations and Measurements of DYSAF-tests, a Research Project to Investigate Safety Limit of Derailment at High Speeds," *Vehicle System Dynamics, Supplement*, Vol. 37, 2002, pp. 543-553.
11. Parena, D., Kuka, N., Masmoudi, W., and Kik, W., "Derailment Simulation, Parametric Study," *Vehicle System Dynamics*, Supplement, Vol. 33, 1999, pp. 155-167.
12. Clementson, J., and Evans, J., "The Use of Dynamics Simulation in the Investigation of Derailment Incidents," *Vehicle System Dynamics* Supplement, Vol. 37, 2002, pp. 338-349.
13. Chen, G., and Jin, X., "On a New Method for Evaluation of Wheel Climb Derailment," IEEE/ASME Joint Railroad Conference paper, Newark, New Jersey, April 4-6, 2000.

APPENDIX C:

Investigation of Wheel Flange Climb Derailment Criteria for Transit Vehicles (Phase II Report)

INVESTIGATION OF WHEEL FLANGE CLIMB DERAILMENT CRITERIA FOR TRANSIT VEHICLES (PHASE II REPORT)

SUMMARY This research investigated wheel flange climb derailment to develop a general flange-climb-distance criterion for transit vehicles in Phase II of the project. The investigations used computer simulations of single wheelsets and representative transit vehicles. The Phase I work investigated the relationships between flange angle, flange length, axle AOA, and distance to climb. Based on these simulations flange-climb-distance equations were developed for some specific wheel profiles.

Based on single wheelset simulation results, Phase II proposed a general flange-climb-distance criterion for transit vehicle wheelsets. The general flange-climb-distance criterion was validated by the flange-climb-distance equations in the Phase I report for each of the wheel profiles with different flange parameters.

Phase II also proposed a biparameter flange-climb-distance criterion for vehicles with an AAR-1B wheel/136-pound rail profile combination. The bilinear characteristics between the transformed climb distance and the two parameters, AOA and lateral-over-vertical (L/V) ratio, were obtained through a nonlinear transformation. The accuracy of the fitting formula was further improved by using a gradual linearization methodology. The biparameter distance criterion based on the simulation results was validated by comparison with the research team's TLV test data. The application to two AAR Chapter XI performance acceptance tests and limitations of the biparameter distance criterion are also presented.

The following conclusions were drawn from the Phase II work:

- A general flange-climb-distance criterion taking the AOA, the maximum flange angle, and flange length as parameters is proposed for transit vehicles:

$$D < \frac{A * B * Len}{AOA + B * Len}$$

where AOA is in mrad and A and B are coefficients that are functions of the maximum flange angle Ang (degrees) and flange length Len (in.):

$$A = \left(\frac{100}{-1.9128 Ang + 146.56} + 3.1\right) * Len - \frac{1}{-0.0092(Ang)^2 + 1.2152 Ang - 39.031} + 1.23$$

$$B = \left(\frac{10}{-21.157 Len + 2.1052} + 0.05\right) * Ang + \frac{10}{0.2688 Len - 0.0266} - 5$$

- The general flange-climb-distance criterion is validated by the flange-climb-distance equations in the Phase I report (Appendix B) for each of the wheel profiles with different flange parameters.
- Application of the general flange-climb-distance criterion to a test of a passenger car with an H-frame truck undergoing Chapter XI tests shows that the criterion is less conservative than the Chapter XI and the 50-msec criteria.
- A biparameter flange-climb-distance criterion, which takes the AOA and the L/V ratio as parameters, was proposed for vehicles with AAR-1B wheel/136-pound rail profile:

$$D < \frac{1}{0.001411 * AOA + (0.0118 * AOA + 0.1155) * L/V - 0.0671}$$

where AOA is in mrad.

- A study of the flange-climb-distance criterion, which takes the friction coefficient as another parameter besides the L/V ratio and the AOA, is recommended for future work.
- The biparameter distance criterion is validated by comparison with TLV test data. Since the running speed of the TLV test was only 0.25 mph, its validation for the biparameter distance criterion is limited. A trial test for validation is recommended.
- Application of the biparameter distance criterion to a test of a passenger car with an H-frame truck undergoing Chapter XI tests shows that the biparameter distance criterion is less conservative than the Chapter XI criteria, including the 50-msec criterion.
- Application of the biparameter distance criterion to an empty tank car derailment test results show that the biparameter distance criterion can be used as a criterion for the safety evaluation of wheel flange climb derailment.

Application limitations of the biparameter distance criterion include the following:

- The L/V ratio in the biparameter distance criterion must be higher than the L/V limit ratio corresponding to the AOA. No flange climb can occur if the L/V ratio is lower than the limit ratio.
- The biparameter distance criterion is obtained by fitting in the bilinear data range where AOA is larger than 5 mrad. It is conservative at AOA less than 5 mrad due to the nonlinear characteristic.
- The biparameter distance criterion was derived based on the simulation results for the AAR-1B wheel on AREMA 136-pound rail. It is only valid for vehicles with this combination of wheel and rail profiles.
- For each of the different wheel profiles listed in Table B-2 of the Phase I report, individual biparameter flange-climb-distance criteria need to be derived based on the simulation results for each wheel and rail profile combination.

CHAPTER 1
A GENERAL FLANGE-CLIMB-DISTANCE CRITERION

The research team investigated wheel flange climb derailment to develop a general flange-climb-distance criterion for transit vehicles in Phase II of the project. The investigations used computer simulations of single wheelsets and representative transit vehicles. The Phase I work investigated the relationships between flange angle, flange length, axle AOA, and distance to climb. Based on these simulations, flange-climb-distance equations were developed for some specific wheel profiles.

Based on single wheelset simulation results, Phase II proposed a general flange-climb-distance criterion for transit vehicle wheelsets. The general flange-climb-distance criterion was validated by the flange-climb-distance equations in Appendix B, the Phase I report, for each of the wheel profiles with different flange parameters.

Flange-climb-distance criteria were developed for each of the rail/wheel profiles, as published in Table B-2 of Appendix B, the Phase I report. Since the wheel and rail profiles vary widely within transit systems, it was desirable to develop a general flange-climb-distance criterion with the maximum flange angle and flange length as parameters for different wheel profiles.

The effects of the maximum flange angle and flange length on climb distance were further analyzed through single wheelset simulations by using 16 wheel profiles with different maximum flange angle and flange length combinations, as listed in Table C-1. The wheel maximum flange angle and flange length were deliberately varied using AutoCAD. The flange root and flange tip were kept the same shape with no restrictions on flange height and thickness.

TABLE C-1 Wheel Profiles Designed by AutoCAD

Maximum Flange Angle Ang (degrees) / Length of Maximum Flange Angle Face $L0$ (in.)	0.252	0.352	0.452	0.552
63 degrees	W1	W2	W3	W4
68 degrees	W5	W6	W7	W8
72 degrees	W9	W10	W11	W12
75 degrees	W13	W14	W15	W16

As shown in Figure C-1, the flange length is defined as the sum of the maximum flange angle length and the flange tip arc length from the maximum flange angle to 26.6 degrees.

Figure C-2 shows the simulation results of these 16 wheel profiles on 115-pound AREMA rail profiles experiencing lateral and vertical forces, which produce an applied L/V ratio of about 1.99. Results are similar to the test (1) and simulation results in the Phase I report and show that the flange-climb distance decreases with increasing AOA. Results also show that the relationship between climb distance D and AOA is nonlinear, with climb distances converging asymptotically to similar values for large AOA.

To develop a general flange-climb-distance criterion with multiple parameters, a methodology was adopted in which the nonlinear relationship between the climb distance and parameters was linearized. This was achieved by using the

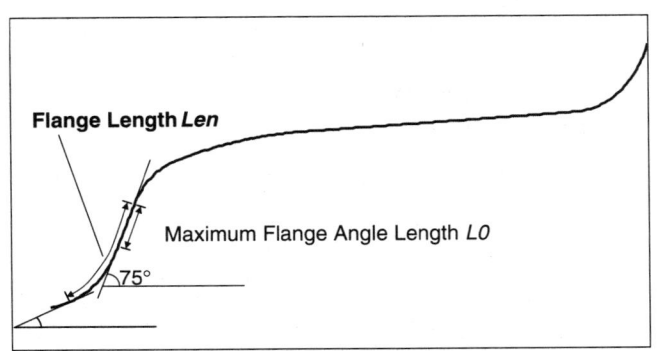

Figure C-1. Definition of the flange length and maximum flange angle length.

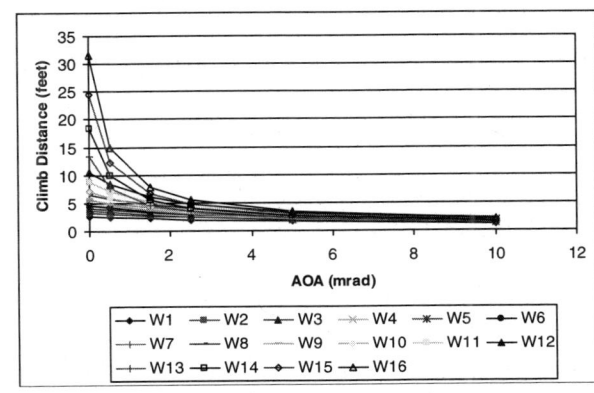

Figure C-2. Effect of AOA on flange-climb distance for different wheel profiles.

following nonlinear transformation function to transform the AOA and distance (D) in Figure C-3 to (x, y) as shown in Figure C-4 for wheelset W1:

$$\begin{cases} X = AOA \\ Y = 1/D \end{cases} \quad (C\text{-}1)$$

The transformed simulation results in Figure C-4 were then fit with high accuracy (R2 of 0.998) in linear form, shown as "fit" in Figure C-4. The linear fit was then transformed back and plotted in Figure C-3.

It is clearly shown in Figure C-4 that the relationship between 1/D and AOA is linear after the nonlinear transformation of Equation C-1. The linear fitting result can be written in the following form:

$$y = ax + b \quad (C\text{-}2)$$

The coefficients a and b for the W1 profile are shown in Figure C-4; i.e., $a = 0.0427$ and $b = 0.3859$. The corresponding nonlinear fitting result shown in Figure C-3 can be written in the following form:

$$D = \frac{m}{AOA + n} \quad (C\text{-}3)$$

where the two coefficients m and n can be calculated as:

$$m = \frac{1}{a}, \quad n = \frac{b}{a}$$

The highly accurate fitting Equation C-3 is obtained, as shown in Figure C-3, due to the benefit of the linear relationship through the transformation.

By using this methodology, 16 formulas were obtained through high accuracy fitting ($R^2 > 0.97$) based on the simulation data at an L/V ratio of 1.99 for each of these wheel profiles listed in Table C-1. Correlation analysis between the two coefficients m and n and the maximum flange angle and flange length were conducted to generate a general function expression.

The coefficient n is decomposed as:

$$n = B * Len$$

where Len is defined as the flange length (in.) from the maximum flange angle Ang to 26.6 degrees as shown in Figure C-1, and B is a coefficient.

Correlation analysis shows that the relation between the coefficient B and the maximum flange angle parameter Ang is roughly linear, as shown in Figure C-5.

Based on the relationship shown in Figure C-5, coefficient B can be expressed in a linear form:

$$B = KB * Ang + CB$$

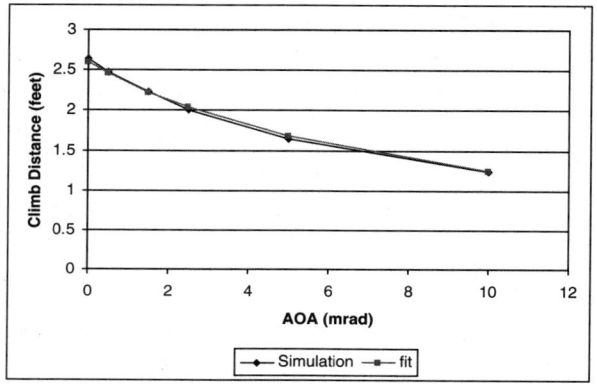

Figure C-3. Effect of AOA on climb distance, W1 profile, 1.99 L/V ratio.

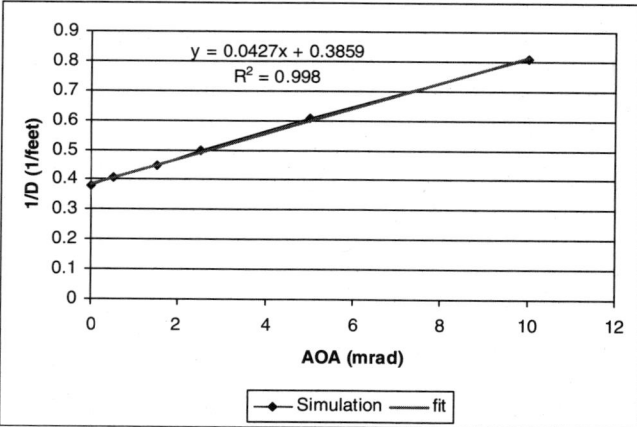

Figure C-4. Linear relation between 1/D and AOA, W1 profile, 1.99 L/V ratio.

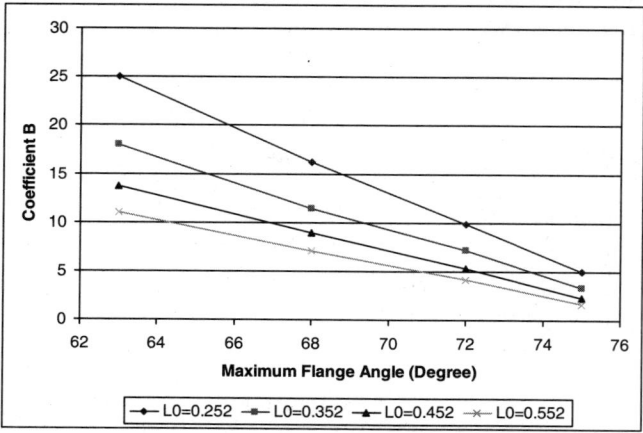

Figure C-5. Effect of maximum flange angle on coefficient B for different wheel profiles, maximum flange angle length L0 in Table 1.

Coefficient *KB* and *CB* are obtained through linear fitting of the lines in Figure C-5. As shown in Figure C-6, the relationship between *KB* and the flange length *Len* is nonlinear.

To get a highly accurate fitting result, the linearization methodology is applied again at this step. First, the nonlinear relationship must be transformed to a linear one. However, no general method was found to construct a transformation function; therefore, a trial and error method was used in this report. This resulted in the following transformation function:

$$\begin{cases} X = Len \\ Y = 10/(KB - 0.05) \end{cases} \quad (C-4)$$

A linear relationship was generated by using this nonlinear transformation function to transform the (*Len, KB*) in Figure C-6 to X, Y. See Figure C-7.

The same methodology is applied to the coefficient *CB* to obtain a linear expression between *CB* and the flange length parameter *Len*.

Coefficient *m* was decomposed as:

$$m = A * B * Len$$

where *B* is another coefficient.

Correlation analysis shows that the relation between the coefficient *A* and the flange angle parameter *Len* is roughly linear, as shown in Figure C-8.

The linearization methodology is used to obtain an expression between the coefficient *A* and flange parameters *Ang, Len*.

Based on the above analysis of the coefficients, a general flange-climb-distance formula with the following AOA and flange parameters is proposed:

$$D = \frac{A * B * Len}{AOA + B * Len} \quad (C-5)$$

where *A* and *B* are coefficients that are functions of maximum flange angle *Ang* (degrees) and flange length *Len* (in.):

$$A = \left(\frac{100}{-1.9128 Ang + 146.56} + 3.1\right) *$$

$$Len - \frac{1}{-0.0092(Ang)^2 + 1.2152 Ang - 39.031} + 1.23$$

$$B = \left(\frac{10}{-21.157 Len + 2.1052} + 0.05\right) *$$

$$Ang + \frac{10}{0.2688 Len - 0.0266} - 5$$

The corresponding general flange-climb-distance criterion is proposed as:

$$D < \frac{A * B * Len}{AOA + B * Len}$$

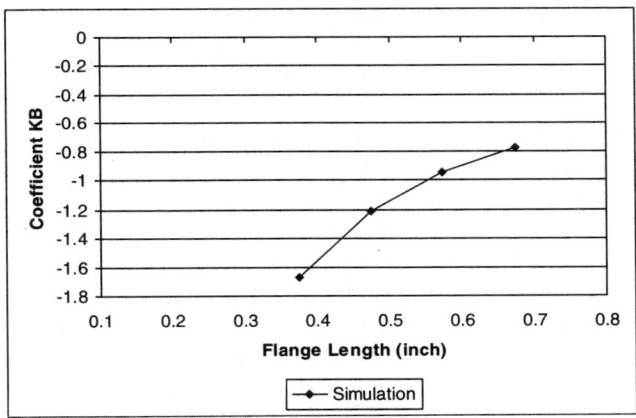

Figure C-6. Nonlinear relationship between KB and the flange length.

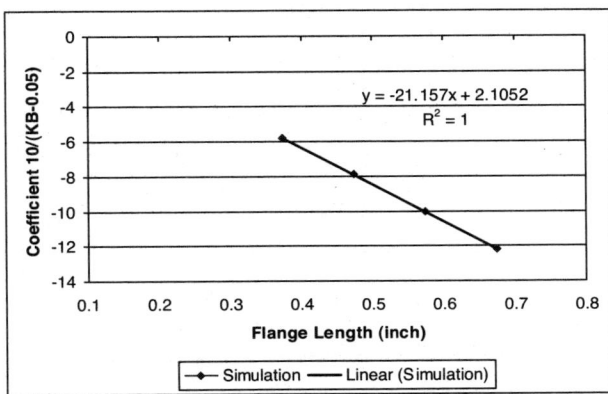

Figure C-7. Linear relationship between the transformed KB and flange length.

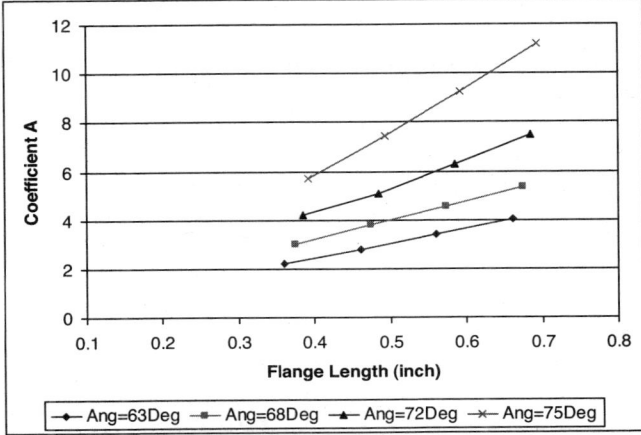

Figure C-8. Effect of flange length on coefficient A for different wheel profiles.

The general criterion was derived from simulation results of the 16-wheel profiles listed in Table C-1. The general equations presented above are considered to be conservative and adequate for use for wheel profiles with flange angles in the normal range of 60 to 75 degrees.

Table C-2 lists a range of limiting flange-climb-distance values computed using Equation C-5 for a specified range of flange angles, flange lengths, and AOAs. Table C-2 indicates that at lower AOA of 5 mrad, the limiting flange-climb distance increases as the wheel flange angle and flange length do. At higher AOA of 10 mrad, flange length has more effect on the distance limit than flange angle.

In summary, considering that flange climb generally occurs at a higher AOA, increasing wheel flange angle can increase the wheel L/V ratio limit required for flange climb, and increasing flange length can increase the limiting flange climb distance.

TABLE C-2 Limiting flange-climb distance computed using Equation C-5

Flane Angle deg	AOA = 5 mrad				AOA = 10 mrad			
Flange Length (inch)	63 deg	68 deg	72 deg	75 deg	63 deg	68 deg	72 deg	75 deg
0.4 inch	2.0	2.2	2.4	2.3	1.5	1.5	1.5	1.9
0.52 inch	2.4	2.6	2.9	2.8	1.8	1.8	1.8	2.1
0.75 inch	3.2	3.5	3.7	4.3	2.3	2.3	2.2	2.4

CHAPTER 2
VALIDATION OF THE GENERAL FLANGE-CLIMB-DISTANCE CRITERION

Six different wheel profiles from several transit systems were analyzed in the Phase I report. Using the maximum flange angle and flange length from these wheels, the relationship between climb distance and AOA was derived from the general Equation C-5 and plotted in Figure C-9. The corresponding climb distance formulas for each wheel profile are shown in the figure.

In the Phase I report, flange climb formulas were developed for these same wheels based on an AOAe for various degrees of curvature. Results were shown in Figure B-32 of the Phase I report and are repeated here in Figure C-10. The shapes of the curves are very similar in nature, with climb distances converging asymptotically to similar values at high AOAs and increasingly sharp curves. The AOAe for transit

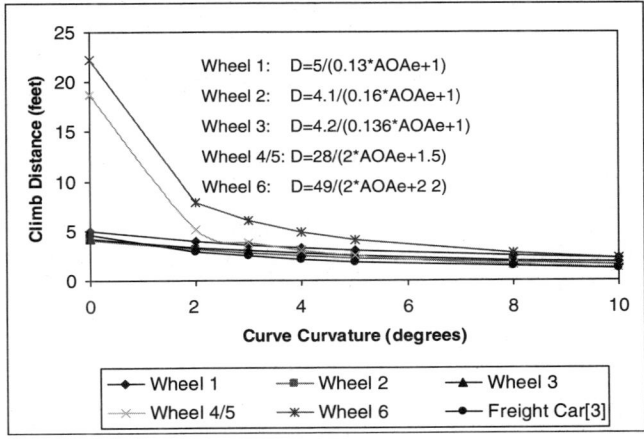

Figure C-10. Climb distance for different wheel profiles.

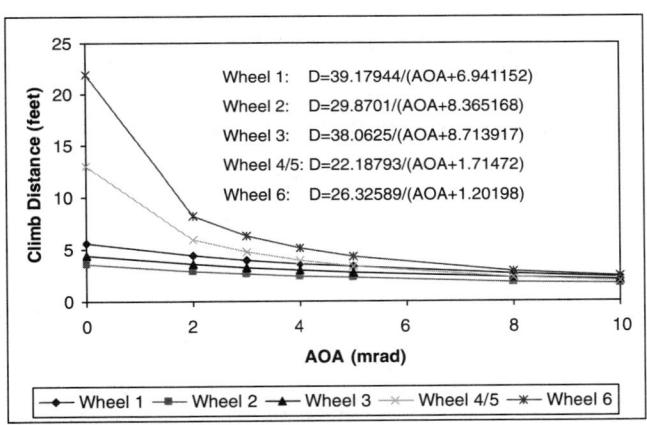

Figure C-9. Climb distance generated from Equation C-5 for different wheel profiles.

and passenger vehicles in curves was derived from the curve radius, based on an assumption that these vehicles do not have significant wheelset misalignments within their trucks and do not have significant wheelset steering angles.

Equation C-5 is derived based on the simulation results when the wheelset was experiencing a 1.99 L/V ratio. As shown in Chapter 7 of this appendix, the average 1.99 L/V ratio (not the peak value) lasting for more than 1 foot is rare according to practical test results. Compared with the measured L/V ratio in practice, the L/V ratio of 1.99 is considered to be conservative enough for transit cars.

The general flange-climb-distance criterion is recommended for use with transit and commuter cars. It is conservative at a lower L/V ratio (< 1.99) and less conservative when the L/V ratio is close to 1.99.

CHAPTER 3

A BIPARAMETER DISTANCE CRITERION FOR FLANGE CLIMB DERAILMENT

The flange-climb-distance criterion proposed in the research team's previous research work (2, 3) for freight cars was based on single-wheelset simulations at a 2.7 L/V ratio for a range of different AOAs. An L/V ratio of 2.7 was considered conservative for freight cars. The general flange-climb-distance criterion in Chapter 2 of this appendix was derived from simulation results at a fixed L/V ratio of 1.99 for different AOAs, which was considered conservative for transit cars. Both criteria were conservative at low L/V ratios, but not conservative enough at L/V ratios higher than the fixed L/V ratio used in the simulations, although the chance of encountering sustained L/V ratios this high is rare in practice. To avoid this dilemma, it is desirable to include the L/V ratio as a variable parameter in the flange-climb-distance criterion.

Results from testing (1) and simulations in the Phase I report show the flange-climb distance decreases with increasing L/V ratio. No flange climb happens (the climb distance is infinite) if the L/V ratio is lower than Nadal's value. Since the L/V ratio is another important factor affecting flange climb besides the AOA, a criterion including the L/V ratio and AOA is expected to reveal more about the physical nature of flange climb and produce more accurate results, although the multivariables fit is more complicated than that of a single variable.

3.1 THE BILINEAR CHARACTERISTIC BETWEEN 1/D AND THE PARAMETER'S AOA AND L/V RATIO

In the following section, a combination of AAR-1B wheels and AREMA 136-pound rail profiles were used in simulations to develop a multivariable fit formula. Figure C-11 shows the simulation results of a single wheelset climbing at different L/V ratios and AOAs.

Figure C-11 shows that the relationship between the climb distance D and the L/V ratio is nonlinear. Through a nonlinear transformation similar to that described in Chapter 1, a linear relationship between 1/D and the L/V ratio was developed (Figure C-12).

Due to the effect of AOA on the creep force, the wheel L/V ratios shown in Figures C-11 and 12 were not the same value for different AOAs even though the same group of lateral and vertical forces was applied to the wheelset. For example, when a 21,700-pound lateral force and 6,000-pound vertical force were applied to the wheelset at different AOAs, the

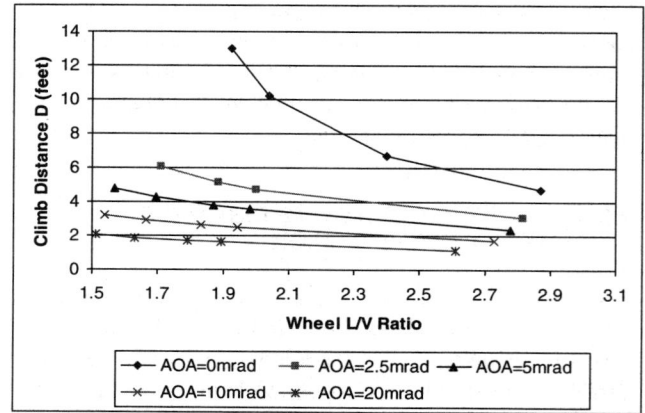

Figure C-11. Effect of L/V ratio at different AOA, 75-degree, AAR-1B wheel, 136-pound rail.

wheel L/V ratios varied with the AOA as shown in the following tabulations:

AOA(mrad)	L/V Ratio (Average value during climb)
0	2.87
2.5	2.82
5	2.78
10	2.73
20	2.61

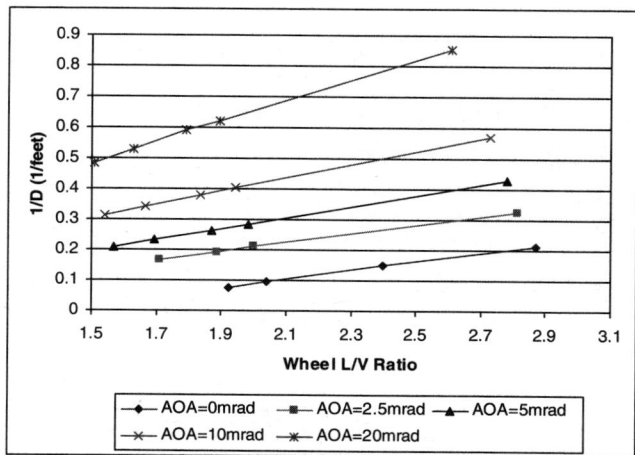

Figure C-12. The linear relationship between 1/D and L/V ratio, AAR-1B wheel, 136-pound rail.

The average L/V ratio for a wheelset being subjected to the same group of lateral and vertical forces at different AOAs, L/Va, for this example is calculated as follows:

$$L/Va = (2.87+2.82+2.78+2.73+2.61)/5 = 2.76$$

The L/Va ratio was used to further describe the relationship between climb distance and AOA for different L/V ratios.

The relationship between the climb distance D and the AOA is nonlinear, as shown in Figure C-13. Again, a similar nonlinear transformation was performed, as described in Chapter 1, with results shown in Figure C-14. The figure shows that there is an approximately linear relationship between 1/D and the AOA higher than 5 mrad. However, it can be seen that the relationship between 1/D and the AOA lower than 5 mrad is nonlinear.

3.2 THE BIPARAMETER CLIMB DISTANCE FORMULA AND CRITERION

Due to the bilinear characteristics between the function of 1/D and the two variables shown in Figures C-13 and C-14, a gradual linearization methodology including two steps described below was developed to obtain an accurate fitting formula. First, the least squares fitting method for two variables was used to fit the simulation result. Since the relationship between the function of 1/D and the L/V ratio is linear for all L/V ratios in the simulations (shown in Figure C-12), the fitting data range for the L/V ratio is the whole data range. But the fitting data range for AOAs is from 5 mrad to 20 mrad to cut off the nonlinear relationship at lower AOAs (< 5 mrad), as shown in Figure C-14. The fitting formula is thus conservative for those AOAs less than 5 mrad, which have a steeper slope than that of the fitting range (5 mrad<AOA<20 mrad). The resulting two-parameter fitting equation in the first step is as follows:

$$1/D = a1*L/V + a2*AOA + a3 \quad (C-6)$$

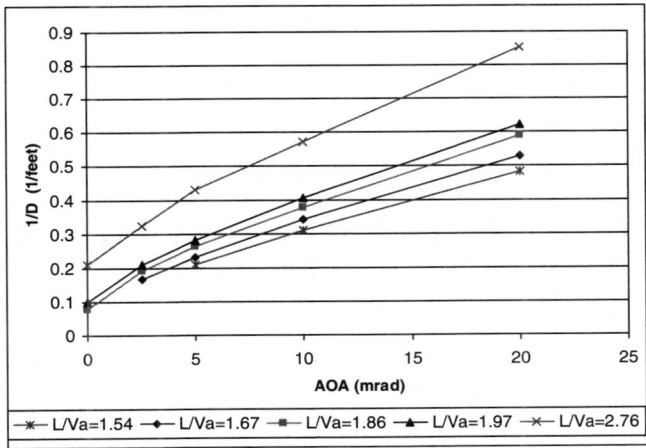

Figure C-14. *Linear relation between 1/D and AOA, AAR-1B, 136-pound rail.*

The fitting accuracy of Equation C-6 may not be satisfactory depending on the simulation model, wheel/rail profile, and the data fitting range. To improve the fitting accuracy, a refinement through further linearization corresponding to the gradual linearization methodology is used in the second step.

The wheelset AOA was kept constant by constraining the axle yaw motions in the simulation. However, the L/V ratio varied during flange climb. The average L/V ratio during flange climb was used in the fitting process. Therefore, in Equation C-6, the coefficient a1 is less accurate than a2 due to the variation of the L/V ratio. Further transformation is performed as follows:

$$Y = 1/D - a2*AOA \quad (C-7)$$

The simulation results were collected as different groups, according to the AOA. For the same AOA simulation group, an accurate fitting equation ($R^2>0.99$) was obtained in the following linear form:

$$Y = b1*L/V + b2 \quad (C-8)$$

The correlation analysis between the coefficient b1, b2, and the AOA for different groups shows that the coefficients b1 and b2 are linear functions of the AOA ($R^2>0.999$):

$$b1 = Kb1*AOA + Cb1 \quad (C-9)$$

$$b2 = Kb2*AOA + Cb2 \quad (C-10)$$

Substituting Equations C-8 to C-10 to Equation C-7, the resulting fitting formula is as follows:

$$D = \frac{1}{[0.001411 * AOA + (0.0118 * AOA + 0.1155) * L/V - 0.0671]} \quad (C-11)$$

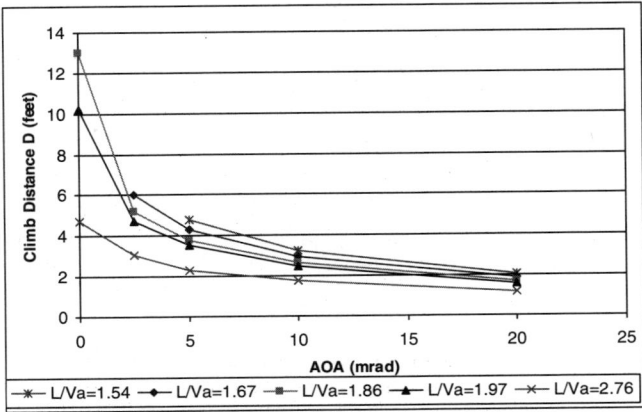

Figure C-13. *Effect of AOA at different L/V ratios, AAR-1B wheel, 136-pound rail.*

Correspondingly, the biparameter flange-climb-distance criterion, which takes the AOA and the L/V ratio as parameters, was proposed for vehicles with AAR-1B wheel/136-pound rail profile:

$$D < \frac{1}{[0.001411 * AOA + (0.0118 * AOA + 0.1155) * L/V - 0.0671]}$$

where AOA is in mrad.

Table C-3 shows the comparison of fitting errors between Equation C-6 and Equation C-11. The fitting accuracy was greatly improved through the "gradual linearization" methodology. The fitting error in Table C-3 is defined as:

$$\text{Fitting Error} = \frac{\text{Formula Value} - \text{Simulation Value}}{\text{Simulation Value}}$$

Based on the above derivation process, some application limitations of the biparameter distance criterion are as follows:

- The L/V ratio in the criterion must be higher than the L/V limit ratio corresponding to the AOA, because no flange climb can occur if the L/V ratio is lower than the limit ratio.
- The biparameter distance criterion is obtained by fitting in the bilinear data range where AOA is larger than 5 mrad. It is conservative at AOA less than 5 mrad due to the nonlinear characteristic.
- The biparameter distance criterion was derived based on the simulation results for the AAR-1B wheel on 136-pound rail. It is only valid for vehicles with this combination of wheel and rail profiles.
- For each of the different wheel profiles listed in Table B-2 of the Phase I report, individual biparameter flange-climb-distance criteria need to be derived based on the simulation results for each wheel and rail profile combination.

TABLE C-3 Fitting errors of Equation C-6 and Equation C-11

Cases	L/VRatio	AOA (mrad)	Fitting Error of Equation C-6 (%)	Gradual Linearization Fitting Error (Equation C-11) (%)
1	1.69	5	20.70	1.58
2	1.87	5	1.68	1.23
3	1.98	5	−8.12	1.31
4	1.67	10	16.91	−1.24
5	1.83	10	1.82	−0.89
6	1.94	10	−6.64	−1.01
7	1.63	20	19.31	0.92
8	1.79	20	4.76	−0.20
9	1.89	20	−1.38	0.97

CHAPTER 4
COMPARISON BETWEEN THE SIMULATION DATA AND THE BIPARAMETER FORMULA

The comparison between the simulation data and Equation C-11 for all L/V ratios at different AOA is shown in Figure C-15. Overall, the results are consistent, especially at AOA greater than 5 mrad.

Figures C-16 through C-20 compare the simulation results with results of Equation C-11 for a range of AOA.

Figures C-16 and C-17 show that Equation C-11 is conservative for AOA less than 5 mrad, with calculated climb distance shorter than the corresponding values from the simulations. Above 5 mrad AOA, the simulations and Equation C-11 match very closely.

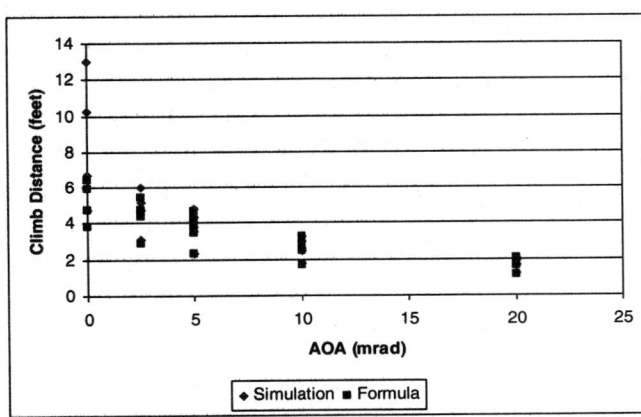

Figure C-15. Comparison between the simulation and equation C-11 for all L/V ratios.

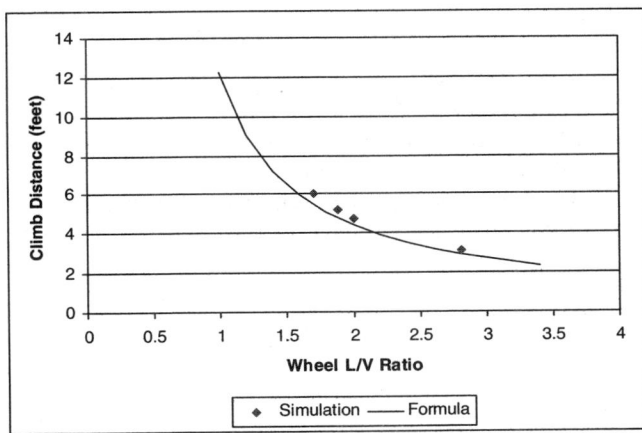

Figure C-17. Comparison between the simulation and equation C-11, AOA = 2.5 mrad.

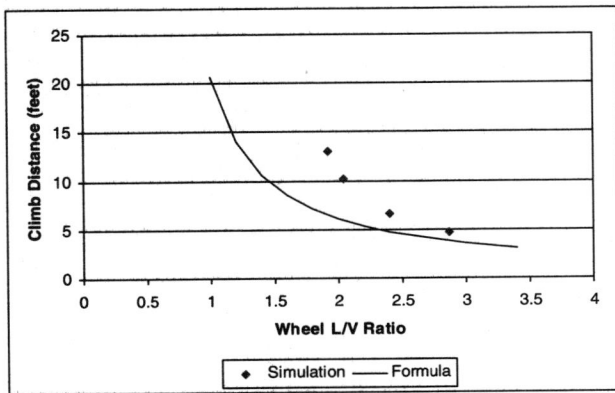

Figure C-16. Comparison between the simulation and equation C-11, AOA = 0 mrad.

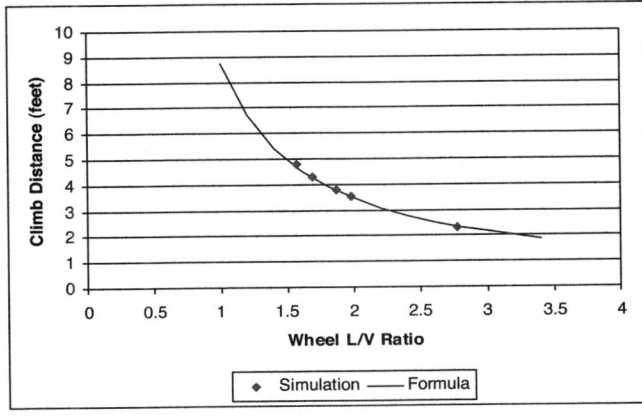

Figure C-18. Comparison between the simulation and equation C-11, AOA = 5 mrad.

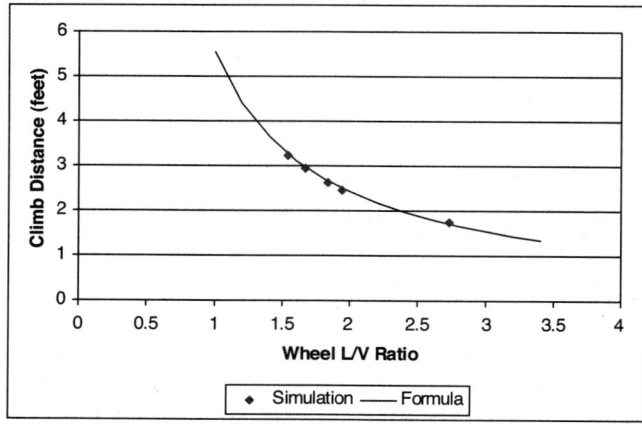

Figure C-19. Comparison between the simulation and equation C-11, AOA = 10 mrad.

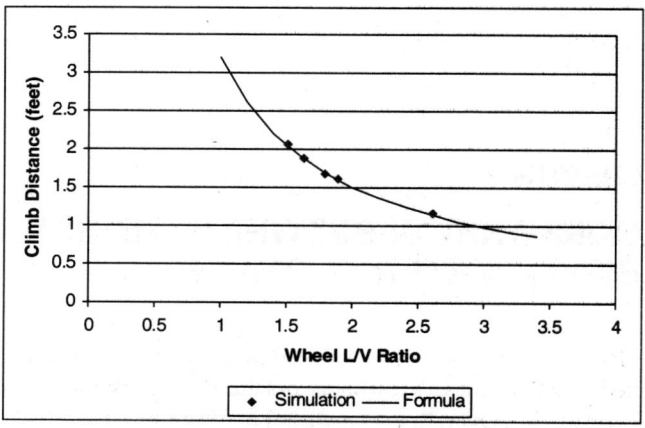

Figure C-20. Comparison between the simulation and equation C-11, AOA = 20 mrad.

CHAPTER 5

VALIDATION THROUGH TLV TEST

The biparameter flange-climb-distance criterion was validated with flange-climb test data from the TLV test on August 25, 1997 (1). The test was conducted on new rails. Since the climb distance is sensitive to AOA, the AOA values were calculated from the test data by the longitudinal displacements (channel ARR and ARL) of sensors installed on the right and left side of the wheelset by using the following equation:

$$AOA = \frac{|ARR| + |ARL|}{93.5} \quad (C-12)$$

where AOA is in mrad and *ARL* and *ARR* are in inches. The distance between the right and left sensor was 93.5 in.

Figure C-21 shows the overall comparison between the test data and Equation C-11 for all L/V ratios at different AOA.

Figures C-21 through C-25 compare the TLV test data with results from Equation C-11 for several of the controlled AOAs. Results of Equation C-11 are more consistent with the test data at higher AOA than at lower AOA.

The difference between the TLV test and Equation C-11, as shown in Figures C-22 through C-25, is due to two main factors: the wheel/rail friction coefficients and the running speed. Equation C-11 was derived based on simulations of a single wheelset with 0.5 friction coefficient at 5 mph running speed. The TLV test was conducted at an average 0.25 mph running speed, and the test data (1) show that the friction

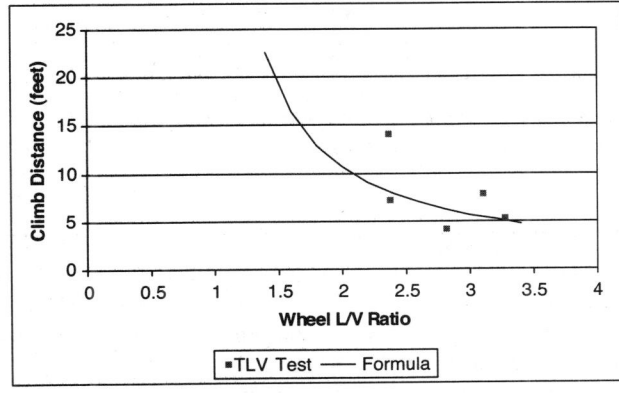

Figure C-22. Comparison between the TLV test and Equation C-11, AOA = -2.8 mrad.

coefficients during test varied from 0.29 to 0.54 for the dry flange face of the new rail.

To demonstrate these differences, three TLV test cases at 32 mrad AOA were simulated by using the single-wheelset flange climb model. The friction coefficients in these simulations were derived from the instrumented wheelset L/V ratios. Simulation results show the L/V ratio converges to Nadal's value when AOA is larger than 10 mrad. For these runs (runs 30, 31, and 32), the L/V ratio just before the wheel climb is

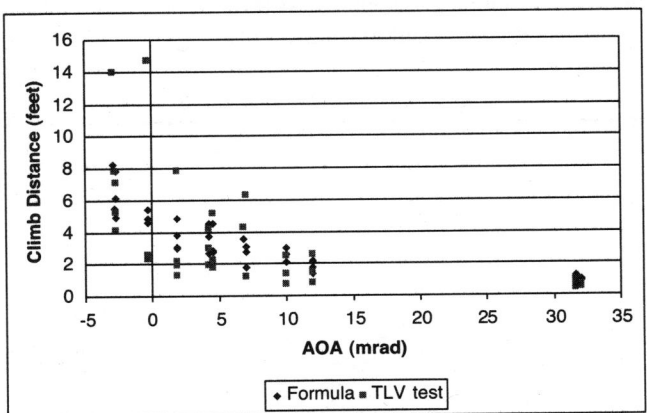

Figure C-21. Comparison between the TLV test and Equation C-11 for all L/V ratios.

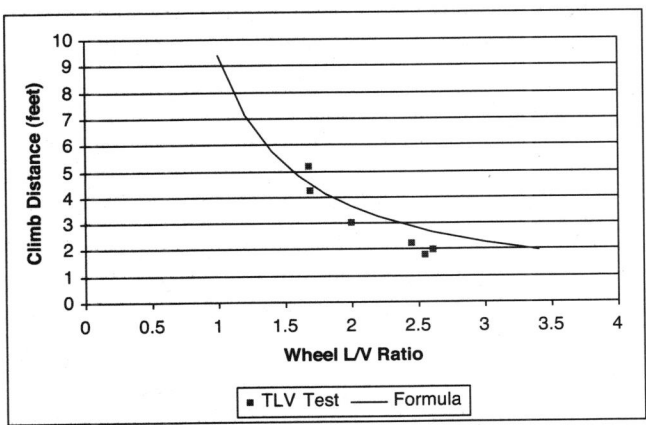

Figure C-23. Comparison between the TLV test and Equation C-11, AOA = 4.4 mrad.

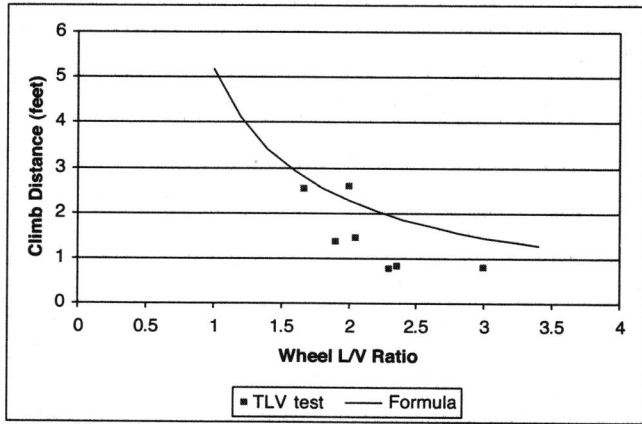

Figure C-24. Comparison between the TLV test and Equation C-11, AOA = 11 mrad.

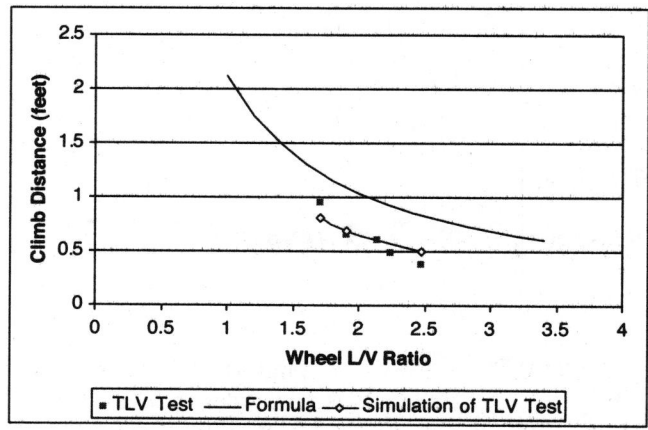

Figure C-25. Comparison between the TLV test and Equation C-11, AOA = 32 mrad.

1.57. The instrumented wheel profile is the 75-degree AAR-1B wheel profile. The friction coefficient between wheel and rail is thus calculated as 0.32, according to Nadal's formula. As can be seen in Figure C-25, the simulations with 0.32 friction coefficient and 0.25-mph running speed show a good agreement between the simulation results and test data.

Considering the running speed in practice, it is reasonable to use a 5-mph simulation speed rather than the actual 0.25-mph test speed for developing the flange climb criteria.

A trend evident in Figures C-22 through C-25 was that the climb distance in the TLV test is shorter than that of Equation C-11 with the increase of AOA. Besides the effect of the lower test speed and the lower friction coefficients in the runs of TLV, the effect of friction coefficients at different AOAs must be considered. In the Phase I report, simulation results show the following:

- For AOAs greater than 5 mrad, the wheel climbed quickly over the maximum flange angle face and took most of the time to climb on the flange tip.
- For AOAs less than 5 mrad, the wheelset took most of the time to climb on the maximum flange angle.

Corresponding to these two situations, the effects of flanging friction coefficients differ:

- For AOA greater than 5 mrad, the climb distance decreases with a decreasing flanging friction coefficient μ because the lateral creep force changes direction on the flange tip to resist the derailment. If μ is smaller, then the resisting force is smaller; thus, the wheelset derails faster than that with a higher friction coefficient.
- For AOA less than or equal to 5 mrad, the climb distance increased with a decreasing flanging friction coefficient μ. The lateral creep force helped the wheel to climb on the flange face and took less time to climb on the tip. In total, it took more time to derail than that with a higher friction coefficient.

The effect of friction coefficients is much more complicated than that of the L/V ratio and the AOA. A study of the flange-climb-distance criterion, which takes the friction coefficient as another parameter besides the L/V ratio and the AOA, is recommended for future work.

CHAPTER 6

ESTIMATION OF AOA

Fixed AOA was used in the single-wheelset flange climb simulations and the TLV test in order to investigate the effect of AOA on flange climb. Both the single-wheelset flange climb simulations and the TLV test have shown that the flange-climb distance is sensitive to AOA. However, the wheelset was not kept at a constant AOA but varied during the climb, as shown in the vehicle simulations. In most practical applications, measurement of instantaneous AOA is not possible. Therefore, to evaluate flange climb potential, an equivalent AOA (AOAe) has to be estimated on the basis of available information (e.g., vehicle type, track geometry, perturbation, suspension parameters) in order to use the biparameter flange-climb-distance criterion in practice.

In the Phase I report, three kinds of representative vehicles corresponding to the Light Rail Vehicle Model 1 (LRV1), Light Rail Model 2 (LRV2), and Heavy Rail Vehicle (HRV) were evaluated. Further simulations, including a freight car with three-piece bogies, were made for these vehicles running on a 10-degree curve, with 4 in. superelevation, and with the AAR Chapter XI Dynamic Curve perturbation. Simulation results were used to estimate the AOAe during wheelset flange climb.

Five running speeds of 12, 19, 24, 28, and 32 mph—corresponding to a 3- and 1.5-in. underbalance and balance (respectively) and a 1.5- and 3-in. overbalance speed—were simulated to find the worst flange climb cases with the longest climb distances.

Longitudinal primary suspension stiffness of the passenger trucks can have a significant effect on axle steering and axle AOA. Therefore, for each of these vehicles, two stiffness variations, which were 50 percent lower and 150 percent higher than that of the designed longitudinal primary stiffness, were used to investigate the effect of suspension parameters on flange climb.

Figure C-26 shows the effect of longitudinal primary suspension stiffness on AOAe, which was calculated as the average AOA during the flange climb.

The warp stiffness of three-piece bogies has an important influence on the AOAe. As shown in Figure C-27, for the worn AAR-1B wheel/136-pound rail profiles, the average AOA during climb decreased with increasing warp stiffness corresponding to the worn truck, new truck, and stiff H-frame truck. For the new wheel/rail profile, the wheel did

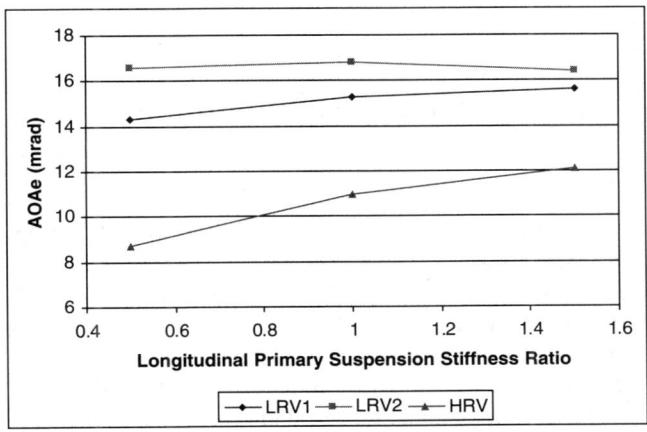

Figure C-26. Effect of longitudinal suspension stiffness on AOAe.

not climb on the rail due to improved steering resulting from the new profile having a larger RRD on the tread than that of the worn profile.

In the Phase I report, an equivalent AOAe formula for the leading axle of a two-axle truck, based on the geometric analysis of the truck geometry in a curve, was derived as

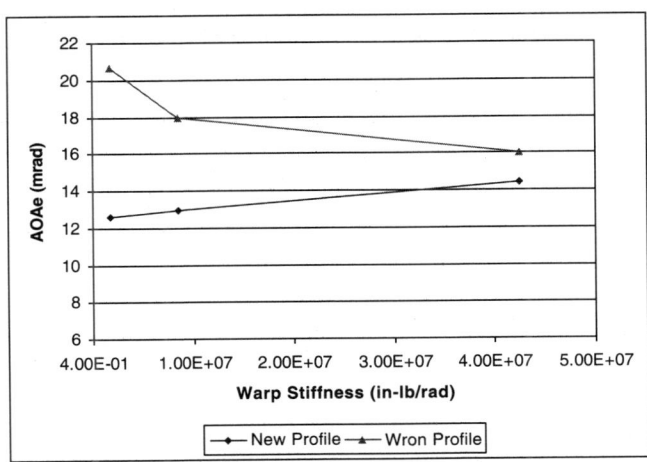

Figure C-27. Effect of warp stiffness on AOAe.

TABLE C-4 Estimation of AOAe

Vehicle Type	Maximum AOAe (mrad)	Axle Spacing Distance (in.)	Constant c
LRV1	16.8	74.8	3.08
LRV2	15.6	75	2.86
HRV	12.1	82	2.04
Freight Car with Three-Piece Bogies (New Bogie)	12.7	70	2.5
Freight Car with Three-Piece Bogies (Worn Bogie)	20.7	70	4.0

Equation C-6. Table C-4 lists the constant c in Equation B-6 of Phase I report (Appendix B) for these four kinds of representative vehicles (LRV1, LRV2, HRV, Three-Piece Bogie) based on the simulation result of the maximum AOAe and axle spacing distance for each of them.

Due to the track perturbations and the degrading of wheelset steering capability, the practical wheelset AOA could be higher than the value calculated by Equation 2.5. The following AOAe, which were considered conservative enough according to the simulation results and test data, were recommended in Table C-5 and shown in Figure C-28.

When the vehicle runs on a curve with the curvature lower than 10 degrees and not listed in Table C-5, it is recommended that a linear interpolation value between the segment points in Table C-5 be used in the criterion, as shown in Figure C-28. Also, it is recommended that AOA statistical data from the wayside monitoring system be used in the criterion to take into account the many factors affecting AOAe if such systems are available.

TABLE C-5 Conservative AOAe for practical use

Vehicle and Truck Type	Straight Lines	5-Degree Curves	10 Degree Curves	Above 10 Degree Curves
Vehicle with Independent Rolling Wheel or Worn Three-Piece Bogies	10	15	20	Equation C-6 (Appendix B) +10
Others	5	10	15	Equation C-6 (Appendix B) +5

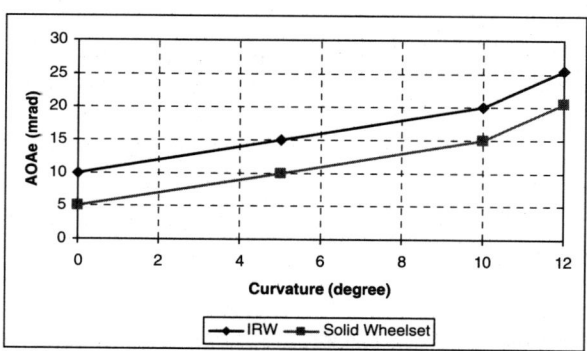

Figure C-28. Recommended conservative AOAe for practical use.

CHAPTER 7

APPLICATION TO VEHICLE DYNAMIC PERFORMANCE ACCEPTANCE TESTS

7.1 APPLICATION TO A PASSENGER CAR TEST

The general flange climb criterion (Equation C-5) and the biparameter distance criterion (Equation C-11) were applied to a passenger car with an H-frame truck undergoing dynamic performance tests at the FRA's Transportation Technology Center, Pueblo, Colorado, on July 28, 1997. The car was running at 20 mph through a 5-degree curve with 2-in. vertical dips on the outside rail of the curve. The L/V ratios were calculated from vertical and lateral forces measured from the instrumented wheelsets on the car. Table C-6 lists the five runs with L/V ratios higher than 1.0, exceeding the AAR Chapter XI flange-climb safety criterion. The rails during the tests were dry, with an estimated friction coefficient of 0.6. The wheel flange angle was 75 degrees, resulting in a corresponding Nadal value of 1.0.

The climb distance and average L/V ratio (L/V ave) in Table C-6 were calculated for each run from the point where the L/V ratio exceeded 1.0. Figure C-29 compares the climb distances to the corresponding distances that are equivalent to a 50-msec time duration. As can be seen, all the climb distances exceeded the 50-msec duration. However, there does not appear to be a direct correlation between test speed and climb duration.

7.1.1 Application of General Flange Climb Criterion

The instrumented wheelset has the AAR-1B wheel profile with a 75.13-degree maximum flange angle and 0.62 in. flange length. By substituting these two parameters into the general flange climb criterion, the flange criterion for the AAR-1B wheel profile is as follows:

$$D < \frac{26.33}{AOAe + 1.2}$$

The axle spacing distance for this passenger car is 102 in., 2.04 was adopted for the constant c since the vehicle and truck design is similar to the heavy rail vehicle in Table C-4. According to Equation B-6 published in the Phase I report (Appendix B), the AOAe is about 7.6 mrad for this passenger H-frame truck on a 5-degree curve. By substituting the AOAe into the above criteria, the safe climb distance without derailment is 3 ft. According to Table C-5, the conservative AOAe for a 5 degree curve should be 10 mrad. The conservative safe climb distance without derailment is 2.4 ft; however, the climb distance according to the 50-ms criterion is 1.4 ft.

The wheel, which climbed a 2 ft distance in the run (rn046) with a 1.01 average L/V ratio (maximum L/V ratio 1.06), was running safely without threat of derailment according to the criterion. The other four runs were unsafe because their climb distances exceeded the criterion.

7.1.2 Application of Biparameters Distance Criterion

Equation C-11 was used to calculate a climb distance criterion for each run, based on the measured L/V ratios, flange angle, and flange length from the test wheels. Because AOA was not measured during the test, the Equation C-11

TABLE C-6 Passenger car test results: climb distance and average L/V (L/V ave) measured from the point where the L/V ratio exceeded 1.0 for friction coefficient of 0.6

Runs	Speed	L/V Maximum	Average L/V	Climb Distance
rn023	20.39 mph	1.79	1.37	6.2 ft
rn025	19.83 mph	2.00	1.43	7 ft
rn045	19.27 mph	1.32	1.10	4 ft
rn046	20.07 mph	1.06	1.01	2 ft
rn047	21.45 mph	1.85	1.47	5.7 ft

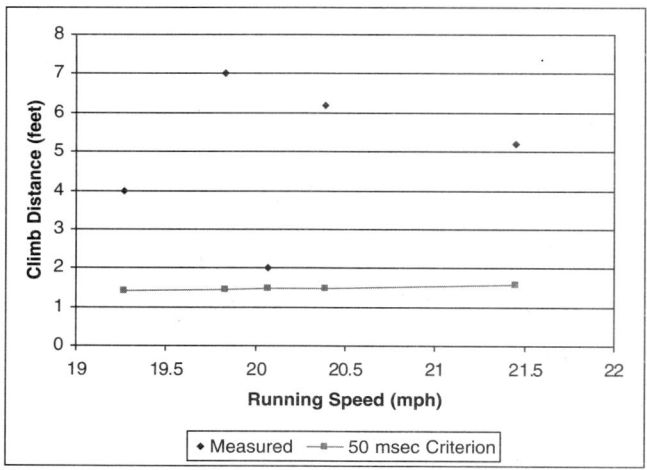

Figure C-29. Application of 50-msec climb-distance criterion.

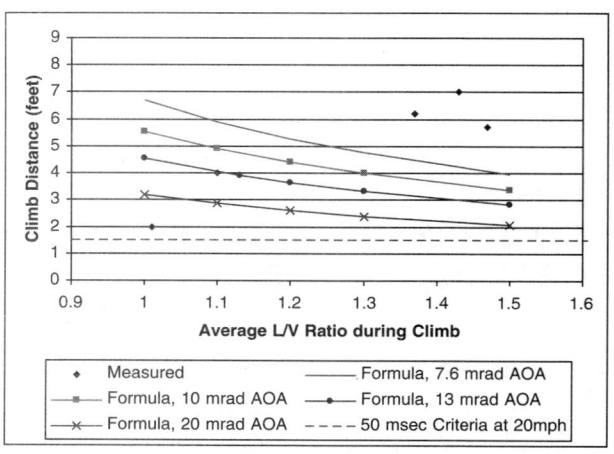

Figure C-30. Comparison of new criterion (Equation C-11) to the 50-msec criterion, 0.6 friction coefficient.

calculation was made for several values of AOA. Results are compared to the 50-msec duration in Figure C-30.

According to the biparameter distance criterion, the run with a 1.01 average L/V ratio (maximum L/V ratio 1.06) was acceptable even for the 20-mrad average AOA, which is an unlikely occurrence for an H-frame truck in a 5-degree curve.

The run with a 1.1 average L/V ratio (maximum L/V ratio 1.32) was acceptable according to the new criterion, as shown in Figure C-30. It would be unacceptable if the AOAe was greater than 13 mrad. This result also means the biparameter distance criterion is less conservative than the general flange-climb-distance criterion.

The other three test runs were unacceptable since they exceeded the new criterion for AOA greater than 7.6 mrad. The same conclusion can also be drawn by applying the criterion with a conservative 10-mrad AOA, according to Table C-5. As noted before, all the test runs exceed the 50-msec criterion.

If a friction coefficient of 0.5 is assumed instead of 0.6, the corresponding climb distances, measured at an L/V ratio higher than Nadal's value of 1.13, are listed in Table C-7.

The run with the maximum L/V ratio 1.06 would then be acceptable because no climb was calculated when the L/V ratio was lower than Nadal's value. The run with the maximum 1.32 L/V ratio would be acceptable since the climb distance was well below the 20-mrad AOAe criterion, as shown in Figure C-31. The other three runs would be considered unacceptable because their climb distances exceeded the 7.6-mrad AOAe criterion line.

The same conclusion can also be drawn if the conservative AOAe (10 mrad) in Table C-5 is used.

This passenger car test shows that Nadal's value, the AAR Chapter XI criterion, and the 50-msec time-based criterion are more conservative than the new distance-based criterion for speeds of around 20 mph. This means that critical L/V values would be permitted for longer distances under the distance-based criterion at low speeds.

7.2 APPLICATION TO AN EMPTY TANK CAR FLANGE CLIMB DERAILMENT

The biparameter distance criterion was applied to an empty tank car flange climb derailment that occurred during dynamic performance testing at TTCI on September 29, 1998. The car was running at 15 mph through the exit spiral of a 12-degree curve. The L/V ratios and wheel/rail contact positions on the tread, measured from the instrumented wheelsets on the car, are

TABLE C-7 Passenger car test results: distance measured from the L/V ratio higher than 1.13 for friction coefficient of 0.5

Runs	Speed	L/V Maximum	Average L/V Ratio	Climb Distance
rn023	20.39 mph	1.79	1.39	5.8 ft
rn025	19.83 mph	2.00	1.45	6.3 ft
rn045	19.27 mph	1.32	1.23	0.7 ft
rn047	21.45 mph	1.85	1.52	5 ft

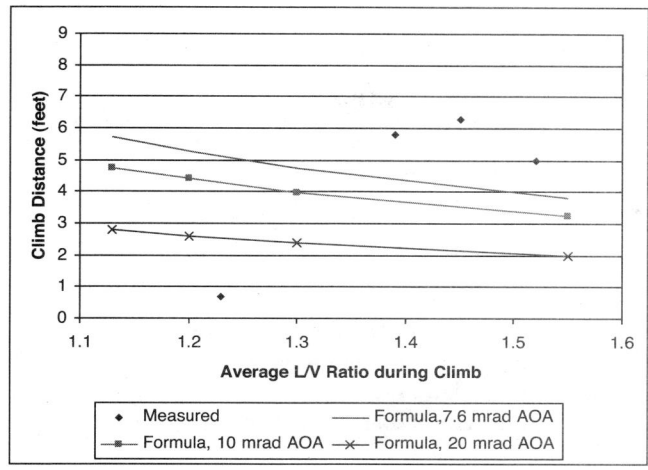

Figure C-31. Application of the new criterion (Equation C-11) for a friction coefficient of 0.5.

shown in Figures C-32 through C-35. Positive contact positions indicate contact on the outside of the wheel taping line, while negative values indicate contact on the flange side of the taping line. Negative values approaching −2.0 indicate hard flange contact. This is shown for Wheel B, which derailed.

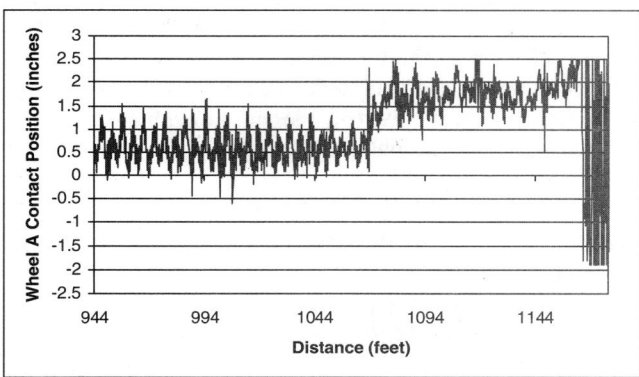

Figure C-32. Contact position on tread of wheel A.

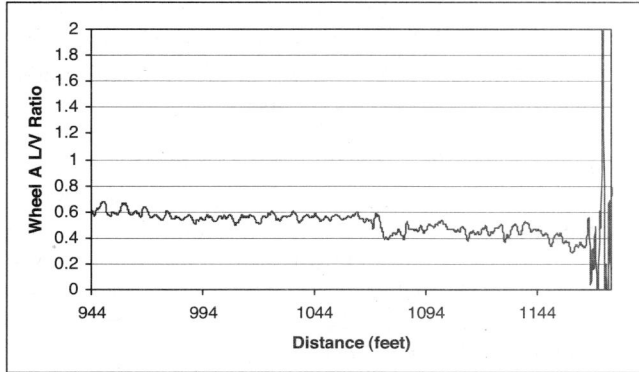

Figure C-33. L/V ratio of wheel A.

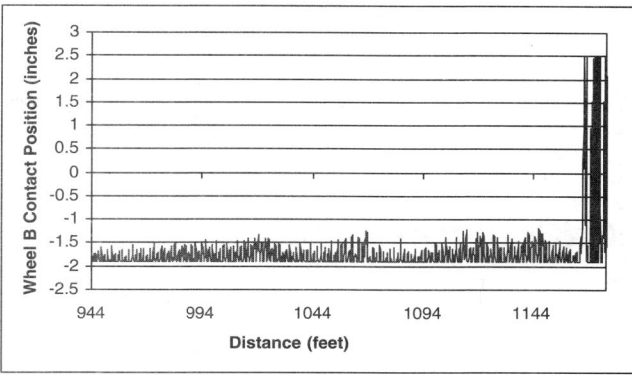

Figure C-34. Contact position on tread of wheel B.

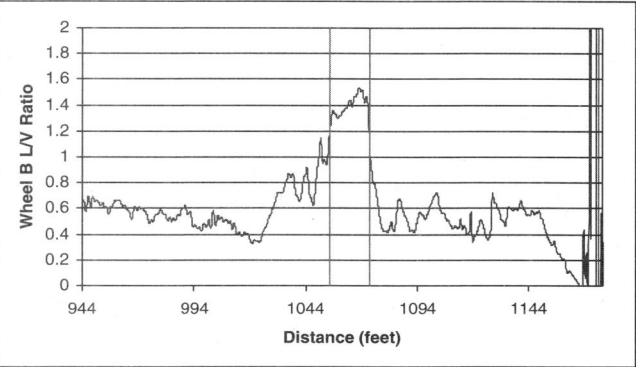

Figure C-35. L/V ratio of wheel B.

The climb distance measured when the L/V ratio was greater than 1.13 (Nadal's value for a 75-degree flange angle and a 0.5 friction coefficient) is 17.9 ft, as shown in Figure C-35. The average L/V ratio is 1.43 during the 17.9 ft climb distance.

The data shown is for an instrumented wheelset that was in the leading position of the truck. The curvature of the spiral during the climb is about 9 degrees. The axle spacing distance for this tank car is 70 in. The constant c was adopted as 2.5, which represents a new bogie in Table C-4. According to Equation B-6 in the Phase I report, the AOAe is about 11 mrad for the three-piece bogie at this location in the spiral curve.

According to Equation C-11, the climb distance is 3.3 ft for the 11-mrad AOAe. The corresponding 50-msec distance at 15 mph would be 1.1 ft. Since the measured climb distance exceeded the value of the biparameter distance criterion, the vehicle was running unsafely at that moment.

Wheel B started climbing at 1,054.6 ft and derailed at 1,164 ft. Therefore, the actual flange-climb distance is longer than 17.9 ft. As shown in Figure C-35, the wheel climbed a longer distance on the flange tip, and the L/V ratio decreased due to the lower flange angle on the tip.

The empty tank car derailment test results show that the biparameter distance criterion can be used as a criterion for the safety evaluation of wheel flange climb derailment.

CHAPTER 8
CONCLUSION

The following findings were made:

- A general flange-climb-distance criterion that uses the AOA, maximum flange angle, and flange length as parameters is proposed for transit vehicles:

$$D < \frac{A * B * Len}{AOA + B * Len}$$

where AOA is in mrad and A and B are coefficients that are functions of maximum flange angle Ang (degrees) and flange length Len (in.):

$$A = \left(\frac{100}{-1.9128 Ang + 146.56} + 3.1\right) *$$

$$Len - \frac{1}{-0.0092(Ang)^2 + 1.2152 Ang - 39.031} + 1.23$$

$$B = \left(\frac{10}{-21.157 Len + 2.1052} + 0.05\right) *$$

$$Ang + \frac{10}{0.2688 Len - 0.0266} - 5$$

- The general flange-climb-distance criterion is validated by the flange-climb-distance equations in the Phase I report for each of the wheel profiles with different flange parameters.
- Application of the general flange-climb-distance criterion to a test of a passenger car with an H-frame truck undergoing Chapter XI tests shows that the criterion is less conservative than the Chapter XI and the 50-msec criteria.
- A biparameter flange-climb-distance criterion, which uses the AOA and the L/V ratio as parameters, was proposed for vehicles with AAR-1B wheel and AREMA 136-pound rail profiles:

$$D < \frac{1}{[0.001411 * AOA + (0.0118 * AOA + 0.1155) * L/V - 0.0671]}$$

where AOA is in mrad.

- A study of the flange-climb-distance criterion that takes the friction coefficients as other parameters besides the L/V ratio and the AOA is recommended for future work.
- The biparameter distance criterion has been validated by the TTCI TLV test data. Since the running speed of the TLV test was only 0.25 mph, one test's validation for the biparameter distance criterion is limited. A trial test to validate the biparameter distance criterion is recommended.
- Application of the biparameter distance criterion to a test of a passenger car with an H-frame truck undergoing Chapter XI tests shows that the criterion is less conservative than the Chapter XI and 50-msec criteria.
- Application of the biparameter distance criterion to an empty tank car derailment test results showed that the criterion can be used in the safety evaluation on the wheel flange climb derailment.

Application limitations of the biparameter distance criterion include the following:

- The L/V ratio in the biparameter distance criterion must be higher than the L/V limit ratio corresponding to the AOA, because no flange climb can occur if the L/V ratio is lower than the limit ratio.
- The biparameter distance criterion is obtained by fitting in the bilinear data range where AOA is larger than 5 mrad. It is conservative at AOAs less than 5 mrad due to the nonlinear characteristic.
- The biparameter distance criterion was derived based on the simulation results for the AAR-1B wheel on AREMA 136-pound rail. It is only valid for vehicles with this combination of wheel and rail profiles.
- For each of the different wheel profiles listed in Table B-2 of the Phase I report, individual biparameter flange-climb-distance criteria must be derived based on the simulation results for each wheel and rail profile combination.

REFERENCES

1. Shust, W.C., Elkins, J., Kalay, S., and El-Sibaie, M., "Wheel-Climb Derailment Tests Using AAR's Track Loading Vehicle," Report R-910, Association of American Railroads, Washington, D.C., December 1997.
2. Wu, H., and Elkins, J., "Investigation of Wheel Flange Climb Derailment Criteria," Report R-931, Association of American Railroads, Washington, D.C., July 1999.
3. Elkins, J., and Wu, H., "New Criteria for Flange Climb Derailment," Proceedings, IEEE/ASME Joint Railroad Conference, Newark, New Jersey, 2000.